The Science of Teaching with Natural Dyes

ISBN: 1-4196-4104-2

To order additional copies, please contact us.
BookSurge, LLC
www.booksurge.com
1-866-308-6235
orders@booksurge.com

The Science of Teaching with Natural Dyes

Jeanne M. Buccigross

2006

The Science of Teaching with Natural Dyes

Contents

To my daughter Lianne Novak and my husband Mike Novak for their help and encouragement. It's been said before, but I would not have been able to complete the project without their help and support throughout the years of research, writing and finally the process of manuscript submission.

In memory of Dr. Arthur Doyle and Dr. John Aber, two talented educators who passed away during the decade that I worked on this book. Dr. Doyle taught a "Physics for Poets" course at my undergraduate college. As a result of that course, I changed my major field from English to Earth Science, and thus changed the course of my life. Many years later, memories of the "Physics for Poets" course inspired me to design and teach a "Science of Art" course and ultimately to write this book. John Aber was a colleague at the College of Mount Saint Joseph and a gifted author of short fiction. John's description of his writing process, specifically editing in afternoons when he couldn't write anything new, helped me make good use of the days when my writing wouldn't flow. In a very real way both of them contributed to this book, although they never knew it.

Foreword

"Why natural dyes?" I have been asked this question many times while writing this book. Initially I began because natural dyes are a topic in my "Science of Art" course at the College of Mount St. Joseph. Many of our art majors are interested in fiber arts and take dye courses using reactive dyes in the Art Department. The text book I use focused mainly on synthetic dyes and the general structure of dyes. It had limited information about natural dyes, however natural dyes proved to be a popular topic for the students' independent projects in the course. Many of these students were not of European ancestry. The natural dyes that were included in the text book were part of a history of dyes and dyeing that focused on European history. It bothered me to be presenting dyes and dyeing as a European accomplishment when I knew that many other cultures were using dyes long before the Europeans. Natural Dyes seemed like a good place to highlight the contributions of women and non-Europeans to the field of chemistry. Naively, I looked for a book or two that would answer my questions about natural dyes, explain how dyes worked, and expand the range of dyestuffs to include non-European cultures.

The first books I collected were recipe books. They answered some questions but raised others. Most were lacking in chemical and historical information. The hobbyist books focused on dyeing wool. My own background as a quilter made me curious about dyeing cotton and other vegetable fibers. I found chemistry books about dyes, but they focused on synthetic dyes and had little information about natural dyes. As I collected and read more and more books what I thought would be a simple topic turned out to be quite complex but at the same time quite fascinating. Like most natural dye hobbyists, somewhere along the line I got "hooked."

At the same time, my multicultural interests expanded beyond the ethnicities of my students to be more global. A well-timed sabbatical leave to Australia allowed the inclusion of Australian and New Zealand dyes in the multicultural survey and the completion of this book. It turns out, not surprisingly, that nearly everybody everywhere used natural dyes for something. What was surprising to me, however, was the fact that peoples from many different cultures and even different hemispheres and continents with vastly different flora were using plants with the very same dye chemicals to do their dyeing.

Meanwhile in my literature survey I found disagreement about nearly everything to do with dyes including the most basic questions about how they worked. Their chemistry is quite complex and there is much about dyes and dyeing that is not well understood even today. As my search became more sophisticated I found (mostly by serendipity) several excellent books with good dyeing advice and good science. It became clear that natural dyes were a good way to teach various levels of chemistry to many students, not just art majors.

There are hundreds of books written about natural dyes and many more about synthetic dyes. I sifted through many of these before I found those most useful for my purposes. My favorites among these are the books by John and Margaret Cannon, James Liles, Rita Buchanan, Rita Androsko and Trudy VanStralen. Jean Carman's book is filled with recipes of eucalyptus dyes if you are lucky enough to live where eucalypts can grow. If you want to buy a few more books, I recommend you consider those.

So, why did I write a book? After spending a decade collecting information from a variety of sources and finally beginning to see a more complete picture, it seemed worthwhile to synthesize it and share it with the chemical education community. Most people wanting to use natural dyes to teach science would want all that information and would have neither the time nor inclination to collect and read all the references I had found. What they would need was practical and useful information at an appropriate level. It also seemed likely that scientifically literate weavers, knitters, quilters and other hobbyists would also be interested in a scientifically accurate but friendly book about natural dyes. Lastly, my experience over the years of writing this book have made it clear that many of my professional scientist friends and colleagues with no interests whatsoever in the fiber arts are also fascinated by the chemistry and biology of natural dyes. This book is the synthesis of all those questions I asked many years ago. In short, it's the book I was looking for when I decided to expand the natural dye section of my Science of Art course. I hope you enjoy reading and using it as much as I have enjoyed researching and writing it.

I owe many thanks to the people who helped me to refine this book, particularly my husband, Dr. Michael Novak. Mike read and edited endless drafts, asked good questions, offered advice on graphics files and helped ensure that my organic chemistry explanations were reasonably accurate, if simplified. Thanks are due to the MSJ students Elizabeth Hess who worked on the cranberry dyes and Tracy Mattingly whose research interests included pH effects of anthocyanin dyes and the importance of co-pigments. She provided me with a good deal of insight into the problems inherent in using natural dyes commercially. Many people helped in the "field tests" of these dyes including a decade of Science of Art students at MSJ. Members of the Oxford Piecemakers Quilting Guild who helped test larger scale dyeing at a natural dye workshop include Debbie Cole, Dixie Woodburn, Conni Kinzler, and Lisa Portwood. Finally thanks are due to my daughter Lianne Novak who helped me test the eucalyptus dye recipes. Lianne gave up part of her school holidays in Australia, somewhat reluctantly, to help me with the experiments. She usually tolerated my writing and reading dye books when she would have preferred to be doing something else. And finally, she tells everyone she knows that her mother is writing a book. Free publicity!

I would like to acknowledge all the people at the College of Mount Saint Joseph that helped along the way. In particular, the students from my Science of Art courses. Their enthusiasm for natural dyes lead me down that path that culminated in the writing of this book. They have taught me as much about art as I have taught them about science. The Faculty Development Committee awarded me a summer research grant that allowed me to start writing a book about natural dyes and, much later, for a sabbatical leave that allowed me to pull it together into a coherent draft and publish it. Thanks are due to former and interim Academic Deans, Dr. Cynthia Zane and Dr. Alan deCourcy for their support of this project. Instructional Media Specialist Brian Bergen spent many hours getting my graphics files into the proper format. Without his help, I would probably still be trying to format figures and tables for this book.

Jeanne Buccigross 2006
Cincinnati and Hamilton OH USA
And Armidale NSW Australia

Part One
A Brief Introduction to Dyes and Dyeing

The development and use of natural coloring agents to dye textiles is one of mankind's earliest endeavors into the world of chemistry (Brunello, Sequin-Frey). The secrets of the early dyers were largely empirical chemistry and even superstition (Sequin-Frey) and some of both remains in the behavior of many natural dyers to this day. The past notwithstanding, natural dyes can be used to illustrate the principles of acid-base chemistry, ionic interactions, metal chelation, hydrogen bonding and nonpolar interactions. They are excellent examples of the importance of molecular orbital theory to adequately describe large organic molecules. They can be used to illustrate the scientific method and the use of appropriate controls. Finally they are an excellent way to show the involvement of women and peoples all over the world in the application of practical chemistry.

Overview of the Book

The topic and use of natural dyes is interesting to science and non-science students alike but is particularly well-suited for courses with art students and other non-science majors, courses that emphasize women's studies and courses with multicultural components. Early American home dyers were women and recipes were exchanged with friends and neighbors (Findley). Plain fabrics and pieced goods were often home dyed in 19th century America. The fastness of dyes was not expected and re-dyeing faded items was common (Brackman). Home dyeing with natural dyes persisted in some parts of the country until well into the 20th century (Waldvogel 1990, Brackman). Dye chemicals such as indigotin, quercetin, quercetagetin, iron oxide, juglone, iron-tannate complexes and barks were used by many different cultures in different parts of the world and from different plants. Some are still being used today.

This synthesizes the practical hobbyist's knowledge of dyes and dyeing with the scientific information available in the reviews. It focuses on dyes suitable for cotton as well as wool. Many fiber artist and quilters use cotton fabrics and will be interested in dyes for their materials. Perhaps more importantly, many traditional cultures used cotton, flax, grasses and other vegetable fibers rather than animal fibers and their dye recipes were developed for these materials rather than wool, silk or other animal fibers.

Part One is introductory material. Chapter 1 provides a brief introduction to the terminology of dyeing and dyes. Chapter 2 includes a brief review of mordants and recipes for their use, as well as instructions on how to properly scour fibers prior to use.

Part Two, Chapters 3-5 is a collection of dye recipes and specific information about each dye. Chapter 3 includes a selection of important dyes used by many cultures around the world. This will be useful to those looking for a multi-cultural approach. Chapter 4 focuses on North American dye plants used by both indigenous peoples of Canada, The United States and Mexico and later immigrants and settlers to the continent. Some popular dyes used by modern hobbyists are included in Chapter 5. This chapter also includes the related craft of flower pounding, which is essentially a dyeing process. All three chapters include the history, the important dye chemicals and practical recipes that have been developed over the years for the Science of Art class at the College of Mount St. Joseph or adapted from another source.

Part Three, Chapters 6-9, includes more detailed chemistry and biology. Chapter 6 explores why we see color and how dyes produce color. Chapter 7 explains how dyes bind to fabrics including the chemistry and action of mordants and assistants. Chapter 8 examines the occurrence of similar dye chemicals in biologically diverse species with the purpose of understanding why so many unrelated plants share common dye chemicals. Chapter 9 explains the processes used to test both light and wash-fastness of dyes in industry.

Finally Part 4, Chapter 10 has several suggested laboratory activities for teachers of various ages of students from elementary school through college and different levels of science instruction. Some of the activities include prepared lab sheets to make it easy for a science teacher to use the activities with his or her class. They may be photocopied for use as is, or modified to meet the particular needs of any classroom. Chapter 11 has suggested references, books and journal articles for additional activities and further background information.

Chapter 1
Before You Begin to Dye

Supplies you will need for dyeing.

If you are a scientist or a high school or college science educator you probably have everything you need at hand. Large Pyrex beakers, hot plates, thermometers, lab gloves, fume hoods, glass stirring rods and the metal salts for mordants make it easy to experiment with the natural dyes or do these activities with your students.

If you are a pre-high school science teacher, a student, crafter or hobbyist, you will need to collect some supplies. Resist the temptation to borrow these things from your kitchen. Natural does not always mean healthy or safe! Remember that arsenic and cyanide are "naturally occurring" compounds. Many plants contain toxic substances in addition to the beautiful dyes. These substances may not be completely removed from a pot or utensil even if it is thoroughly cleaned. Don't risk poisoning yourself or your family by using your cooking supplies as dyeing supplies. Likewise, it is not safe to dye in your kitchen unless you are dyeing with a food product. Many of the dye plants produce disagreeable odors and sometimes potentially toxic vapors while extracting the dye. Use a hot plate, camp stove or barbeque grill out in your yard to ensure adequate ventilation.

Large un-chipped enamel pots make excellent containers for dyebaths and can be purchased inexpensively. Stainless steel pots are more resistant to damage but are usually more expensive. If you will be mordanting only with alum, then aluminum pots may also be used. Buy very large pots if you intend to dye fabrics or yarns for sewing or crafts rather than samples for chemistry experiments. What looks like a big pot in the store will quickly look like a small pot once you have added fabric. Wire strainers are useful for straining out dyestuffs. Coffee filters can be used but will drain more slowly and will retain some of the dye chemical. Wooden spoons make good stirrers although they will become permanently stained with the dye and glass stirring rods may be more useful for older students. Gloves are a necessity to keep your hands from turning colors at the very least, and to protect yourself from potentially toxic or caustic mordants and dye chemicals. If you want to precisely control the temperatures then you will need thermometers. Remember - anything you use for dyeing must be dedicated to dyeing and never used for cooking! Dedicated dye pots can often be collected inexpensively at garage or yard sales or at bulk or seconds shops.

When you are starting out, enamel pots are probably the cheapest and easiest route. If you get "hooked" then investing in some "bargain basement" stainless steel pans will make the job easier.

Chemicals used in dyeing can be purchased from many sources. If you are a scientist or science teacher, you can easily obtain all the supplies you need. If not, you will have to shop around a bit more. Convenient but sometimes expensive sources for washing soda, mordants and some dyestuffs are the supermarket or pharmacy. Garden and swimming pool suppliers will also have some of the mordants and assistants available. Chemical suppliers will generally have most of the items available for lower cost. Specialty dye suppliers can also be found. The best sources of all these items will vary depending on your situation and your geographical location. Looking in the telephone book or doing an internet search are the best ways to find suppliers in your area.

What is a Natural Dye?

Before you begin to dye it is best to familiarize yourself with some commonly encountered dyeing terms. Table 1 summarizes the dyeing terms used in this book and additional detail for some terms is given in this chapter. A more complete understanding of their meanings will come as you read through the book but this section of the book will give you sufficient background to begin.

Table 1 The "top ten list" of dyeing terms used in this book. A more complete understanding of their meanings will come as you read through the book, but this will give a good basis to begin your reading. This brief list is not alphabetical, but rather arranged in appropriate order to give a basic understanding of dyes and dyeing.

Dye - a colored substance or mixture of substances that is dissolved in solution when applied to the fiber. The dye permanently attaches to a fiber or fabric through a chemical interaction or physical entrapment process and colors the fabric.

Natural dye – a dye that is found in nature or is derived from plant, animal or mineral sources by extracting or fermenting a naturally occurring precursor.

Dye chemical – the actual chemical coloring agent(s). In natural dyes "dye chemical" is not synonymous with the "dye" because the dye often contains many other natural products. The dye chemical may be thought of as the "active ingredient" of the dye. In a synthetic dye, the two terms are synonymous in most cases.

Fast – a dye that stands up to light and washing with little or no fading. This is a relative term as no natural dyes are as fast as modern synthetic dyes. The opposite of a fast dye is a fugitive dye.

Dyestuff – usually a plant or part of a plant that contains the dye and can be used to produce a dye. Some animal or mineral substances can be dyestuffs. Most of the dyes in this book are vegetable dyes.

Dyebath – the solution of dye after it has been extracted from the dyestuff and is ready to be used for dyeing the fabric or fibers.

Afterbath – as the name implies is a treatment used after the fiber has been dyed. This may be treatment with a mordant, an acid or base to adjust the pH, or another dye

Mordant – a chemical substance, often a metal ion, that is required for a dye to be fast.

Cellulosic or Vegetable Fiber – fibers or fabrics derived from plant materials. These include cotton, linen, flax, jute, ramie, grasses, reeds and so forth.

Dyes are defined as different from stains and pigments based on the way they interact with fibers. In general, as defined in Table 1, a dye is considered as a colored substance that is dissolved in solution when applied and then permanently attaches to a fiber or fabric through a chemical interaction or physical entrapment process. Although in general stains are also applied as solutions, Liles defines staining as involving no "chemical combination" with the fiber or fabric. This distinction is somewhat problematic in our every day use of the words dye. Many books and people talk about dyeing basketry materials (presumably reeds and grasses) and these are chemically similar to cotton and other vegetable fibers, however the process of application is different and the role of physical entrapment is probably minimal while it can be quite important in dyeing cotton fibers. Early Hawiians "dyed" tapa, a bark cloth, in secret ritualized procedures (Krohn). Similarly we talk about staining of wood but wood and cotton chemically are both primarily cellulose, and wooden baskets are often dyed with cotton dyes.

Pigments are somewhat easier to distinguish from dyes and stains in most cases. A pigment may be chemically identical to a dye chemical, but the method of application is different. Pigments are used in paints. Unlike dyes and stains they are not in solution, but are applied as a suspension of fine powder in a carrier like water or oil. Pigments adhere to the surface of the paper or fabric by virtue of a "binder" or glue-like substance also present in the paint. While the differences between dyes and pigments are more clear cut than that between dyes and stains, there is some overlap when pigments and paints are used with fabrics. Similarly some dyes enter the fabric in solution but are oxidized to become insoluble after they enter the fabric. These insoluble dyes often have no "chemical combination" with the fabric although the often cited definition of a dye includes a chemical interaction with the fiber (Liles).

Perhaps the best definition of a dye is that afforded by the Color Pigment Manufacturers Association cited by Lewis. In their system pigments and dyes differ in their form when *applied* to the object rather than their chemical composition. Their definition requires that a pigment be crystalline (insoluble) during the entire coloring process. On the other hand, a dye must be dissolved when it penetrate the fabric, after which it may or may not remain in a water soluble state (Lewis).

Iron oxide is a good example of a chemical that is often classified as all three. Iron buff dyeing is a process of applying a soluble iron oxide to fabric for the purpose of changing its color. Ochre is iron oxide that is frequently used as a pigment in paints, but some cultures decorated fabrics and fibers with ochre and with other mineral pigments also found in cave paintings (Gerber). Clearly this does not alter the chemical definition of a dye and a pigment given above, but does it make ochre a dye in the everyday sense of the word dye?

Dyes can be characterized by the nature of the interactions between the dyes and the fibers and this is quite useful for synthethic dyes. Most natural dyes fall into the categories of acid and acid-mordant dyes however (Aspland, Storey), so those classifications are not particularly useful with natural dyes. Instead natural dyes are often grouped into broad categories based on their chemical structures, for example anthocyanins, flavonols, naphthoquinones etc. These distinctions will be useful for the reader using natural dyes to teach organic chemistry or botany to science students, however readers teaching more basic science will not need that level of detail. In this book a few such categories will be briefly discussed because of the large number of dyes in the category. For readers interested in more detail, these organic structural categories have been described elsewhere (Goodwin & Mercer, Sequin-Frey).

Most practitioners of natural dyeing divide natural dyes into more pragmatic categories. Two categories are based on the way the dyes can be used. Dyes are classified as "direct" or "substantive" dyes if they will make a fast dye without pre-treatment of the fiber. They are called "mordant" or "adjective" dyes if they require pre-treatment or simultaneous treatment with another substance called a "mordant" (see Table 1 and Ch. 2) to produce a fast color. The categories are based on a dye's *ability* to act as a direct dye even if it is not used as one. Direct dyes are often used with mordanted fabric either to modify or intensify the color of the resulting dye, but they are still classified as direct dyes. Indeed most direct dyes act as direct dyes because the dyestuff contains not only the dye chemical but significant concentrations of some natural mordant.

Direct dyes attach to the fibers without need of any chemical bridge or mordant. Simplistically, direct dyes may be described as binding to cellulosic or vegetable fibers (cotton, flax, reeds etc) by hydrogen bonding to the many hydroxyl groups on cellulose, by hydrophobic/nonpolar interactions with the fiber and by physical entrapment of the large dye molecule in the fibers. This is somewhat simplistic, as most "direct dyes" are produced from dyestuffs that contain significant amounts of tannin and the tannin acts as a mordant at least with cotton. The binding to protein or animal fibers (wool, silk) conversely is by ionic interactions between charged groups on the dyes and charged amino acid side chains of the wool and silk proteins as well as nonpolar interactions. All these processes are in fact quite complex and not completely understood. They are simplified here as an introduction but are described more completely in Ch 7.

Mordant dyes are dyes that will not produce a fast dye unless the fabric is treated with a mordant, usually a metal salt. The most commonly used mordants are the cations of Al, Cu, Cr, Fe, Sn. Most authors consider tannins or tannic acid to be a mordant as well. Some consider anything used in the dyeing process such as cream of tartar or tartaric acid and Glauber's salts to be mordants as well while others refer to these as "assistants" or "other chemicals." In simple terms, the metal ion mordants appear to work by forming complexes with both the dye and the fabric. Tannins seem to work by binding with the dye and the fiber molecules. These processes are described in more detail in Ch. 7.

Record Keeping

Dyeing with natural dyes will never be perfectly reproducible. Even if the exact same procedure is followed in exactly the same way, there will be variations in the colors obtained. This is primarily due to natural variations in the dyestuff. The amount of dye chemical in a dyestuff varies with season, weather, soil type and geographical location. It may even vary from plant to plant in the same garden. The dyer who expects to get exactly the same results each time will be disappointed and frustrated. On the other hand, a dyer who accepts that variations from batch to batch are part of the charm of using natural dyes, will be open to natural variations and will rarely be disappointed.

Even with natural variations in mind however, you can maintain some control over your dyeing process if you follow the directions carefully and if you use consistent technique from batch to batch. Dyeing with natural dyes is not "rocket science" but good experimental technique can minimize the variation obtained when dyeing and will guard against dye failures. This will be particularly true if using purchased dyestuffs. Commercial dyestuffs will generally vary less than dyestuffs collected in the field or grown in a home garden.

If it is important to you to control your results as much as possible, then keep a detailed notebook of your procedures and weigh all dyestuffs and fabrics. The notebook should also include comments, any abnormalities and a small sample of the dyed material. Always be sure to note the source of the dyestuff, whether it was used fresh or dried, the length of time the dyestuff and the fabrics were simmered, whether a mordant was used and what type of fiber was used, such as bleached cotton, unbleached muslin etc. All these factors will affect the final colors to a marked degree. Liles has achieved very good reproducibility by conducting his dyeing as a scientific exercise. VanStralen likewise carefully monitors and controls her procedures to make her dyeing as reproducible as possible.

Chapter 2
Preparing Fabrics for Dyeing

Before dyeing your fabrics or fibers, they must be properly prepared. In general this requires a careful cleaning called scouring followed by a mordanting process. For even dyeing, the fabric must be wetted before dyeing. Fabrics and fibers can be dyed and will take up dye if they have been laundered without fabric softener even if they have not been scoured, however the colors may not be as deep or as fast (Liles). For a teaching laboratory exercise this might be acceptable. The procedures for scouring vary depending on the fabric being used and from author to author. This book will recommend an intermediate level of scoring, but if you are concerned about thorough scouring to ensure the best possible results for dye uptake and color-fastness a more rigorous procedure is available in Liles book.

Most dyes are not permanent unless mordants are used. There are a few exceptions to this, but in general mordanting with alum for wool or alum/tannin for cotton is recommended for the recipes in this book. An iron mordant is suggested for a few specific dyes, but in general the less toxic alum is a good choice for dyes that will involve children or adults unfamiliar with good chemical techniques and safety. It also makes disposal of the spent mordant solutions easier. Alum is used in making pickles and in swimming pools so small amounts of alum can safely be poured down the drain or around acid loving plants in the garden (Liles). This is not true for most other mordants and proper disposal will vary depending upon where you live. You should contact local authorities for advice on disposal of heavy metal mordants.

Introduction to Traditional Mordants

When following published dye recipes from hobby dye books, knowing what chemical to use as a mordant and how much to use is complicated by the frequent use of archaic common names for chemicals. Recipes and lists are regularly copied from other sources, usually with no attributions. The chemical names used in one source may be incorrectly transcribed in others, so that it is difficult to know what actual metal salt or organic compound was used by any author. This makes it particularly difficult to be sure you are using the same chemical in the proper amount, particularly when the weights of mordant are given.

Alum is commonly used as a mordant with cotton and other cellulosic fibers, generally in conjunction with tannins or tannic acid and sometimes cream of tartar and/or washing soda (Buchanan, Cannon & Cannon, Carman, Hallett, Kraemer, Liles, Schetsky). Alum is quite frequently used with animal fibers as well. Several authors refer to "two different alums," namely Potassium Aluminum Sulfate and Ammonium Aluminum Sulfate (Buchanan, Cannon & Cannon, Casselman) and indeed one author refers to the first as "poisonous alum" and the second as "nonpoisonous household alum" (Casselman) although neither is very toxic. Cannon & Cannon refer to these as "potash alum" and "ammonia alum" but the Merck Index lists only potassium aluminum sulfate under the index name of alum (Merck) and this appears to be the most widely used as a mordant. In either case, it appears that the important ion is the Al(III) and the recommended amount of alum varies so widely that errors due to molecular weight differences between the two salts are small by comparison. Merck does not list any toxicity for alum, but it is generally considered to be the safest or the mordants. Ammonium alum is used in pickling and first aid (Casselman) and potassium alum in swimming pools (Liles) as well as dyeing. Aside from the practical issues of safety, alum is also the most important historical mordant (Ponting). Alum is the only mordant that will be recommended in this book with the exception of the historical iron and iron-tannate dyes.

Blue vitriol or bluestone is the old name given for copper mordant and most authors simply refer to this as "copper sulfate" (Buchanan, Cannon & Cannon, Casselman, Lesch, and van Stralen) however the Merck Index and Adrosko refer to blue vitriol as copper sulfate pentahydrate (Merck, Androsko). From the descriptions in the recipe books such as a blue powder or a gardening supply, most of the authors appear to be using the pentahydrate and simply referring to it as copper sulfate, so weights given in most references probably are for the pentahydrate compound. Some dyers use copper pots or add copper scrubbing pads to the dye bath instead of adding the mordant (Buchanan, Glasson & Glasson) although others dispute whether sufficient metal enters the dyebath to sufficiently mordant the fabric (Liles). Copper mordants are not required for the recipes in this book. They can be used with students who have been trained in proper chemical safety techniques to expand the range of colors given by a dye.

Copperas, green vitriol and iron vitriol are old names given for the iron mordant. Most authors simply refer to this as ferrous sulfate more correctly named iron (II) sulfate (Cannon & Cannon, Casselman, Lesch, van Stralen). The Merck Index gives ferrous sulfate heptahydrate as a synonym for copperas as do Androsko and Liles (Merck, Androsko, Liles). Here it appears that some authors are in fact using the heptahydrate while others may be using the anhydrous salt, but all referring to it simply as ferrous sulfate. Van Stralen, for example, shows a beige powder as her ferrous sulfate, which is probably the anhydrous salt while others refer to its blue-green color (Carman 1978), which indicates the heptahydrate is being used (Merck). This makes interpreting the weights given in recipes difficult.

In spite of the universal reference to *ferrous* sulfate, it is most likely that the actual mordant is Fe(III) and not Fe(II). Aqueous solutions of Fe(II) sulfate oxidize rapidly to Fe(III) when heated (Merck) and most dye and mordant recipes require heating and simmering of the mordanting solutions so it is unlikely that significant Fe(II) persists in the solution. The rate of oxidation is increased in alkaline solutions or with exposure to light (Merck). In fact, The Merck Index lists iron (III) sulfate as a mordant rather than iron (II) sulfate. In spite of this it appears likely that most authors are using some form of iron (II) sulfate. As with copper, some authors refer to using iron pots or steel wool pads as a way to introduce iron to a dyebath (Glasson & Glasson) particularly if it is to be used as an afterbath rather than the mordant. Nearly all authors say that use of iron pots or utensils will sadden or dull the dye color. Liles cautions against the use of chipped enamel pans because they are iron underneath the enamel and will affect the dye color (Liles).

Iron mordant is required for a few of the recipes in this book. Iron can be toxic particularly to children, even in small amounts. The recipes using iron mordants should only be used by students trained in proper chemical safety and technique. Teachers of younger children may choose to dye samples of fabric with the iron mordants to show students rather than having the students dye the fibers themselves.

Chrome mordant is referred to by most authors as potassium dichromate (Adrosko, Buchanan, Cannon & Cannon, Casselman, and van Stralen) but Lesch says there is more than one kind of chrome, and refers to the "potash dichromate" (potassium dichromate) as being most commonly used and working the best. Both potassium dichromate and sodium dichromate dihydrate are listed as "chrome" in the Dictionary of Dyes and Dyeing (Ponting). The Merck Index does not list "chrome" as a synonym for potassium dichromate, but does list mordant among the uses for dichromate. The physical descriptions of the mordants in the books make it likely that in fact all are using potassium dichromate. In the chrome mordanting of wool, Cr (VI) of the dichromate oxidizes methionine, cysteine and tyrosine amino acid residues in wool and is simultaneously reduced to Cr (III) (Christie, Maclaren & Milligan, Trotman). The Cr(III) is then bound by carboxyl groups on the wool (Maclaren & Milligan) and coordinates dyes (Christie, Trotman). Thus Cr (III) is the active form of the chrome mordant.

Dichromate is highly toxic (Merck Index). Readers are strongly cautioned against the use of chrome mordant. It should **never** be used by children under any circumstances. It should not be used by any teachers or students who have not had significant chemistry training. Disposal of the spent solutions is also a problem as they cannot be safely disposed of except by hazardous waste professionals. No recipes in this book require the use of chrome mordant. While it is often used today in commercial dyeing (Trotman), chrome was not used as a mordant prior to 1800 (Ponting) and thus has limited historical value. Students can experience the same type of dye chemistry with the much safer alum mordant. I never use chrome as a mordant with or without students as I believe the risks outweigh the benefit of a slightly larger range of colors from a dye.

Tin mordant is generally listed as stannous chloride (Buchanan, Cannon & Cannon, Casselman, Lesch, and van Stralen). The Merck Index lists both stannous and stannic chloride as mordants (Merck). Stannous chloride forms an insoluble hydroxide in neutral aqueous solutions but is soluble in acid and basic solutions. It is also a powerful reducing agent (Merck) but what role this plays in the dye process is not clear. Tin mordants are not required for the recipes in this book.

Other Mordants and Assistants

Tannins (Sequin-Frey, Lesch, Androsko) or tannic acid (Cannon & Cannon, van Stralen) are referred to by some authors as mordants and by others as assistants. Tannins are necessary for a fast dye with cotton (Cannon & Cannon, Carman 1978, Liles) and in fact many of the substantive dyes for cotton are produced from dye stuffs which contain the dye as well as significant amounts of tannin. These include onion skins (Cannon & Cannon), tea (Hendrickson), black walnuts (Liles) and eucalypts (Glasson & Glasson).

Cream of Tartar is also referred to as tartaric acid by some authors (van Stralen) but most refer to it as potassium hydrogen tartrate (Androsko, Buchanan, Cannon & Cannon, Casselman, Ponting, Schetsky) and this agrees with the Merck Index usage (Merck). Some authors say that this is equivalent to the "cream or tartar" available for cooking use (Casselman, Schetsky) however Cannon & Cannon say that household "cream of tartar" is not tartrate. This must be an error on their part as food guides also identify "cream of tartar" as potassium hydrogen tartrate (Food decoder). Cream of tartar is generally used to regulate the pH of the dyebath.

Glauber's salts are used by some dyers to "level the dye bath" or to "standardize the dye." Those who refer to these salts by any other name simply list sodium sulfate (Cannon & Cannon, Casselman) with the exception of Ponting who calls it sodium sulfide. This is certainly an error as the Merck Index lists Glauber's salts as a synonym for sodium sulfate decahydrate (Merck). The use of Glauber's salts is quite common in the older dye references (Lesch) but is less common in the newer books. The ionic strength of a dyebath has been shown to affect the uptake of dye sometimes in a complex manner (Aspland, Motomura *et. al.*). Uneven dyeing is a problem for very large dyes that bind very strongly to fibers and less problematic with smaller dyes with lower affinities for fabric (Christie) such as natural dyes. Most authors now do not use Glauber's salts in their dyeing but do recommend the use of deionized water, distilled water, rain water or at the very least soft water whenever possible. The recipes here do not use Glauber's salts.

Scouring recipes

There are many literature methods for scouring fibers and they vary greatly in details including the temperature, length of time, amount and type of detergent. The recipes included here are intermediate versions. Both an easy washing machine scouring method and better hand scouring methods are provided. The intended use of your fiber and the amount of fiber you need to prepare should be considered when choosing a method. I have had good luck on most occasions with the washing machine methods. I nearly always dye using purchased cotton and wool yarns and fabrics which have been cleaned to some extent already. Liles reports that many dye failures are due to improperly scoured fibers including purchased fiber, and that this is particularly true with cotton. My personal experience has also been that inadequate washing will result in very little dye uptake. It is important to plan the dyeing day ahead of time so the fabric can be properly prepared. The gentler scouring methods are written for 120 g (about ¼ lb) of fabric or fiber. The washing machine methods can easily scour several yards of fabric at a time, and are a better choice for children's t-shirt dyeing days and similar projects.

Easy Wool Scouring Method

This method is recommended for fabric rather than wool yarn, although yarn can be cleaned by this method to be used for student samples. In that case tie the yarn in loose hanks and enclose them in a net lingerie bag. Yarn for any other purpose than samples should be cleaned by one of the gentle methods below.

Choose a laundry detergent specifically advertised for washing wool and use the amount of detergent recommended by the manufacturer. Set the washing machine on the warm temperature setting. Choose the shortest time on the gentle/hand wash cycle on the machine to minimize the agitation then stop the machine toward the end of the wash cycle and let the wool soak for 30-60 minutes before the rinse. Rinse in warm water. Follow with an extra rinse in cooler water. Do not use fabric softener or any other additive. This method is particularly good for large amounts of fabric but care must be taken not to felt the wool by excessive agitation or sudden temperature changes. It is the method I generally use to prepare yarn or fabrics for my science classes to dye.

If using a washing machine that does not have a gentle cycle or the manufacturer does not recommend laundering hand washables, a modification of this method can be used. Set the temperature as above and fill the machine but shut it off before it agitates. Add the wool and soak as above. Drain the water but do not let the wool spin. Remove the wool and fill the machine for the rinse cycle. Again add the wool and soak it without running the machine. Drain again. Fabric should not be dried in a clothes dryer.

Gentle Wool Scouring- Method I

Dissolve ½ -1 tsp. mild detergent designed for wool in 5 L of warm water. Soak the wool for several hours. Squeeze out suds. Rinse in water of the same temperature that the bath was when the wool was removed. If using anything but a pH neutral detergent, the wool should be rinsed in a vinegar bath then rinsed again in clear water. This method is good for commercially prepared yarns. It should not be used for fleece. Method II is safe for fleece.

Gentle Wool Scouring – Method II (adapted from Thompson & Thompson)

Dissolve 1.8 mL (scant ½ tsp) of mild detergent in hand- hot* water. Let fleece soak overnight, or yarn for 1 hour. Prepare a new bath of hot water and enough soap or detergent to produce a good lather. Place the fleece or yarn in the new bath and soak for an additional hour. Remove and rinse several times. It is important to use rinse water of the same temperature as the bath to prevent felting. Alternately, the wool can be rinsed in successively cooler rinses as long as sudden temperature changes are avoided. The wool can be partially dried by gentle spinning in a "salad spinner" and then left to air dry. Never use a clothes dryer.

* So called "hand hot" water (40-50°C) can be judged without a thermometer because it will feel hot to your hand (Wickens). This method may be used for fleece.

Easy Cotton Scouring Method

Choose a laundry detergent advertised as being tough on stains, particularly grease and greasy stains. Use the amount of detergent recommended by the manufacturer for heavy soil. Washing soda (sodium carbonate) may also be added following the directions on the package, however I do not generally use it. Put the fabric and detergent in the washing machine set on the hot water temperature setting. Use a higher water level than you would normally use to launder the same amount of clothing so the fabric can move about freely. Cotton yarn should be enclosed in a net lingerie bag and washed on a gentle cycle to keep it from tangling. Wash the fiber on a long cycle and if possible stop the machine toward the end of the wash cycle and let it soak for 30-60 minutes before the rinse. Rinse in warm water. Follow with an extra rinse or rerun the washer on warm temperature setting without adding any additional detergent. Do not use fabric softener or any other additive. The fabric can be used wet or dried in a clothes dryer for storage or later use.

This method is particularly good for large amounts of fabric. It is the method I generally use to prepare yarn or fabrics for my science classes to dye or for dye workshops with my quilting guild. It would be an appropriate method to wash t-shirts or other clothing items to be dyed in a workshop with children.

Better, More Rigorous Cotton Scouring (adapted from Liles)

For 120 g of fabric dissolve 1 Tbs. washing soda and 1 tsp. laundry detergent in 6 L of water in a large beaker, or enamel or stainless steel pot. Add fabric or yarn and let soak at least 30 min-1 hr. After soaking bring to a boil and boil 1 hour. Rinse thoroughly in hot water and then in cooler water. This method is most practical for small amounts of fabric or fiber. I seldom use this method unless I want to make sure the fabric will dye very well and I have a small sample to be dyed. Liles contends that many dye failures, particularly with cotton, are due to inadequate scouring. If the sample is important, this method is the safest bet.

Wetting the Fiber or Fabric

Dry fabric or fiber will not take up mordant or dye evenly and will be mottled or spotty after dyeing. Often it will not be dyed as deeply as wetted fabric. If you start with fiber straight from the scouring bath that is still wet it may be mordanted with no further treatment. If you have dried the fabric it needs to be wetted before mordanting. Similarly freshly mordanted fibers do not need to be wetted before beginning dyeing, but mordanted fibers that have been dried will need to be wetted. Tepid water should be used for wool to prevent felting but hot water should be used for cotton and other cellulosic fibers. To wet a fiber for mordanting or dyeing simply submerge it in water and let stand for 30 min. (Thompson & Thompson). Squeeze out excess water and immediately mordant or dye.

Mordanting Recipes

There is less variation in alum mordanting recipes than in scouring recipes, however there are several common methods. For vegetable fibers, both a faster one-day method and a more rigorous 2-3 day method are given. The longer method is recommended if fabric is to be used for craft or clothing, while the faster method may be appropriate for student labs. Even the 1 day method requires soaking overnight, so mordanting of vegetable fibers must be done at least one day in advance of dyeing. For wool all the methods are very similar and require only a few hours. All that differs is the amount of alum the different authors recommend. All methods given are for 120 g of fabric or fiber, which corresponds to approximately ¼ lb.

Cotton mordanting methods can generally be used for any vegetable fibers. The recipes require the use of tannic acid. If tannic acid is unavailable, tannins from other sources can be used with the following equivalencies given by Liles: 1 g tannic acid = 2 g tara powder = 4 g dry sumac leaves = 8 g fresh sumac leaves = 10 g oak galls (Liles). Most recipes require 7 or 8 g of tannic acid for 120 g of fiber so the equivalencies can be easily calculated by students either with simple ratios or using the equivalencies as "conversion factors."

The processes for wool mordanting are shorter and simpler than for cotton. With the exception of Carman, they are nearly identical differing only in the amount of alum and cream of tartar recommended. The range reported for alum is between 17 g and 30 g with most being in the 20's so an intermediate value of 25 g is used here. The recommendations for cream of tartar are less varied with 7 g being most common. The recipes provided here are for fabric or yarns not for fleece. Methods for fleece dyeing and mordanting are similar but fleece must be treated more gently to prevent felting. Carman 1978 reports that cream of tartar is not necessary when dyeing with eucalypt dyes and she mordants her wool using the recipe for cotton but simmers the wool in the mordant bath for 30 min. rather than 1 hr. Carman's wool method is not recommended for dyes other than eucalyptus.

One Day Alum Mordant for Cotton and Vegetable Fibers
(adapted from methods of Carman and Wickens)

25 g Alum (Potassium Aluminate Sulfate)
7 g Tannic Acid*
7 g Washing soda (Sodium carbonate)
4 L Distilled or soft water

Dissolve the alum and tannin in a small amount of hot water. Heat 4 L of water in large beaker, stainless steel or enamel pan to nearly boiling. Stir in the dissolved alum/tannin mixture. Add wet cotton and boil for 1 hour. Let fabric stand in the bath overnight. Rinse and use immediately or dry for later use.

*Carman 1984 does not use tannic acid but she dyes using eucalyptus leaves that are known to contain large amounts of tannin. She does not rinse the mordanted fabric and recommends dyeing immediately rather than drying for later use. It is not clear how much of the alum will be retained in the fiber if it is rinsed or dried before use because cotton does not retain metal ions (Androsko, Liles).

Two or Three Day Method for Cotton*
(adapted from methods by Androsko, Buchanan, Kraemer, Liles, Schetsky)

60 g alum (Potassium aluminum sulfate)
8 g Washing soda (Sodium carbonate) **

8 g Tannic acid

Day 1: Dissolve tannic acid in 4L hot (50°-80°C) distilled or soft water. Add wet cotton and work a few minutes, then sink and let stand overnight.

Day 2: Dissolve alum in 500 mL of hot (50°-80°C) distilled or soft water then let cool to room temperature. Dissolve washing soda in 100 mL of distilled or soft water and slowly add to alum bath while stirring. Bubbles of carbon dioxide gas will be produced. When these subside add room temperature water to bring the bath to 4 L volume. Add the wet fiber and work for a few minutes then let stand overnight.

*Only Liles method uses a two day process. Most authors of the three day method divide the alum and washing soda in half and prepare two 4L alum baths (Androsko, Kraemer, Schetsky) with Day 1 the first alum bath, Day 2 the tannin and Day 3 the remaining alum bath. These methods also boil the fabric in the alum bath for 1-2 hr. before letting stand overnight. Buchanan uses a three day method where Day 1 is the full strength alum bath, Day 2 the tannic acid bath and Day 3 a second soak in the same alum bath. She does not simmer her mordanting baths. As the tannins are needed to retain the alum in the fabric (Androsko, Liles) the two day method is probably more appropriate.

**Buchanan does not use washing soda in her alum bath. Most cotton mordanting recipes do require the use of washing soda. The alkaline pH appears to be more effective in mordanting vegetable fibers (Liles).

Alum Mordanting of Wool
(adapted from methods of Buchanan, Cannon & Cannon, Castino, Hallett, Kraemer, Liles)

25 g Alum (Potassium aluminum sulfate)
7 g Cream of tartar
4 L distilled or soft water

Add room temperature water to a large beaker, stainless steel or enamel pan. Dissolve the alum and cream of tartar in a small amount of boiling water and add this to the bath. Add wet wool and slowly raise the temperature to a simmer over a half hour to forty five minute period. Simmer the wool an additional hour. Let cool in bath to lukewarm. Remove and rinse with same temperature rinse water to prevent felting.

Iron mordant for Black Walnut Black Dye (adapted from Androsko, Kraemer and Liles)

4 g iron (II) sulfate•heptahydrate
4 g cream or tartar
4 L distilled or soft water

Dissolve the iron in a small quantity of hot water then dilute to 4L with warm water in a large beaker or stainless steel or enamel pot. Add wetted cotton and work for several minutes then allow to stand overnight. OR

Add wetted cotton to a 0.01 M $FeSO_4$ solution and let stand. This is more concentrated than the 1g/L (3.6 mM) mordant bath suggested above. It is included as an option because many chemistry stockrooms will have iron solutions of this concentration.

Preserving Dyestuffs and Dyebaths

Some dyestuffs must be used fresh while others can be successfully dried or frozen. In most cases this information is given in the dye recipes if I have successfully preserved the dyestuff or have read of its preservation from a reliable source. I routinely dry coreopsis, marigold, weld, mint, purple basil, bronze fennel and other herbs and flowers that I grow for dyeing. Sometimes the colors are identical to the fresh materials and other times slightly different. I have successfully frozen black walnut hulls for long periods of time with no ill effect. Although I have not tried it, Castino reports great success in storing prepared dyebaths rather than the raw dyestuff by freezing for up to 3 months (Castino). This is an intriguing idea for dyestuffs that do not dry well and are best processed when fresh. Most dyebaths cannot be kept at room temperature for any length of time without development of mold. The interested reader is encouraged to experiment with Castino's method of freezing prepared dyebaths.

Part Two
Dyes and Dye Recipes

There are hundreds or perhaps thousands of natural dye plants world wide. Some dyes, such as indigo, are found in large numbers of closely related species. Other dyes are unique to one particular subfamily or even species. This section is a discussion of some of those dyes and the plants that produce them. It includes basic recipes that can be easily used in a home or school setting. The dyes have been grouped into categories to make it easier for a teacher to choose a dye palette to be used as part of an integrated curriculum, however there is a great deal of overlap among the three categories and some dyes could easily be placed in more than one category.

In most case, the amount of fiber that can be dyed varies somewhat and depends on the depth of color desired. The general and widely accepted "guidelines" are given in Table 2 (Cannon & Cannon, Carman). These should be used as a starting point. The actual results may well vary from plant to plant. Some plants contain more dye chemical than others. In general if very saturated colors are preferred, then are higher dyestuff:fiber ratio will be needed. If less saturated, pastel colors are preferred then a smaller dyestuff:fiber ratio will work.

If in doubt, it is better to use more rather than less water. In most cases, if sufficient dye stuff is used a more concentrated dyebath will not produce a darker fabric but will result in uneven dyeing caused by fiber crowding. This seems contrary to common sense, but it occurs because the binding of dye to fiber is an adsorption process. Given sufficient dye molecules and sufficient time dye uptake will continue until the fabric has bound dye molecules to all the available sites on the fiber. This is not dependent on the amount of water in the dyebath, but only the ratio of dyestuff to fiber. Unless otherwise noted in the recipe, the amounts of dyestuff given here are suitable for 100 g-120 g (about ¼ lb.) of fiber or fabric. The corresponding yardage will vary greatly depending on the fabric, but in general is between ½ and 1 yard (or 1 meter) of fabric. Paler shades will be obtained if more fabric is used or if the dyebath is re-used for additional fibers.

Table 2 The generally recommended ratios of dyestuff to fiber and water for various plant materials. These should be considered as guidelines only. See text for further information.

Type of material	Mass plant material	Mass fiber	Volume water
Flowers, dried	25 g	25 g	1 L
Flowers, fresh	50 g	25 g	1 L
Fruit	25 g	25 g	1 L
Leaves	50 g	25 g	1 L
Bark	25 g – 100 g	25 g	2 L
Roots	25 g – 50 g	25 g	1 L

Chapter 3
A Multicultural Selection of Dyes

A great variety of plant and mineral substances have been used as dyes and dyestuffs by different cultures. Some dyes and dye chemicals, however, were used throughout the world by many different cultures. Sometimes these come from similar mineral or plant species, as is the case with the iron buff. Other times the same chemical is found in a variety of plant families as is the case with the dye chemicals indigotin, quercetin, quercetagetin and the iron-tannate complexes. Six dye chemicals are presented here along with information about where and when they have been used, what plants or minerals contain them and how to find them. There are some dyes in the other sections that were also used by more than one culture, but they were not as universal as those presented here. Each dye in this multicultural section was used on at least three different continents.

Iron Buff Red, Orange or Yellow

Staining with iron oxide is one of the oldest methods used to dye fabrics, paint the body and paint other items of human interest (Brunello, Liles). It was used all over the world (Brunello, Gerber). The items to be dyed were often simply buried in mud (Brunello, Liles). Iron buff dyes in the form of mud, clay, ochre or iron stained rain water were used by Native Americans including the Nez Perce of the Columbia Plateau in the northwest U.S. (Harris), the Navajo (Bryan & Young) and Hopi (Colton) of the southwest U.S., the Pima Indians and other Latin America natives (Harris). It was also used by Western African tribes (Stuart Robinson), European cutures including Russians (Brunello), the Japanese (Harris), early American settlers (Liles) and Southern African-American families through the Depression era (Ramsey & Waldvogel).

In recent historical times, iron buff dye was intentionally produced from the chemical copperas (iron sulfate) or iron pieces in vinegar (aqueous iron acetate), and used by American colonials (Bassett & Larkin, Liles), and emancipated African-American slaves and their descendents (Waldvogel '94). Iron buff dye was even used commercially in the early American fabric printing industry. Fabrics were first impregnated with iron acetate and then treated with an alkaline solution to produce iron oxide on the fabric (Bide). Australian Aboriginals and the Indo people of Mato Grosso used iron buff from ochre as body paint (Brunello, Lloyd) and the Eskimo peoples painted their weapons with iron oxides (Brunello).

Iron buff dye is still used commercially by Paradise Sportswear Inc. of Hawaii to produce their "red dirt"™ shirts and tote bags. Paradise Sportswear has returned to the traditional method of burying their shirts and cotton tote bags in a "secret mixture" of red Hawaiin mud and organic matter. Photo 1 shows a purchased "red dirt" tote bag.

Simple Iron Buff Recipe (adapted from Androsko):

Prepare an iron solution by dissolving 45 g of iron (II) sulfate heptahydrate in 1 L of water. Prepare a soap bath by dissolving 45 g of powdered laundry detergent soap in 1 L of water. First, place wet cotton in the iron solution and agitate for a few minutes. Remove and squeeze out water, then dip into the soap solution and work for a few minutes. Remove and squeeze out soap. Repeat the dipping process 3 or more times until desired shade is reached.

Passive Iron Buff Recipe

Prepare an iron solution by dissolving 45 g of iron (II) sulfate heptahydrate in 1 L of water. Place wet cotton in the solution and let stand for hours or days. You will get a pale shade of orange or yellow-orange, probably not as deep as the recipe above. The color might be improved slightly by using a soap afterbath.

Slow Iron Liquor Recipe:

Place several small iron nails or iron filings into a glass or plastic container and cover with vinegar to a depth of 1-2 in. above the nails or filings. Keep covered to reduce evaporation and let stand for several weeks or even months. The longer the mixture stands, the more iron acetate you will produce. Once the solution is sufficiently orange-brown it is ready. Remove any remaining nails or filings by straining the solution through a wire mesh and retain the "liquor." Fabric samples can be dyed by immersing wet fiber into the solution and letting it stand for a week or more. Although this is a very slow method, it is worth trying to see how patient our ancestors were when dying with iron buff.

"Get Down and Dirty" Iron Mud Recipe:

50 g of iron rich soil*
25 g distilled water or rain water
6 in square (2 g) bleached combed cotton or cotton yarn

Put the mud in a plastic container and add the wet cotton. Agitate the cotton until it is completely covered. Cover the container to prevent the mud from drying out but periodically open the container and stir the mud to oxygenate the bath. Let stand for at least 1 week. The longer it stands the better. This is a fun recipe, although it is messy and is best done outside.

*The natural iron content in soil varies from 0.5% to 80% (Frye). The best way to find iron rich soil is to enlist the help of a geologist or agricultural extension office. Iron rich soils often appear reddish orange but may also be darker.

Iron-Tannate Black and Slate Gray

Black dyes produced from iron-tannates are found around the world (Brunello) and were used in ancient times to dye hides (Stuart Robinson) and other garments (Liles). Sumac, Hemlock or galls were a common source of tannins (Liles). Both the Navajo and the Hopi use a similar ancient recipe with sumac, *Rhus trilobata*, pinon pitch, *Pinus edulis* and iron oxide heated in the fire to produce black dye (Bryan & Young, Colton). The Tzotzil people of Mexico produced a black dye by boiling fibers with an iron rich soil (Sayer). American colonists produced a gray dye using an iron mordant and sumac without heating in the fire (Bronson) and some hobbyists still use this today (Van Stralen). Oriental carpets from Turkey are often dyed with iron-tannate dyes (Gallagher).

The Maori people of New Zealand make a traditional black dye for flax by first treating the flax with a mordant made from one of three local tree barks makomako, *Aristoltelia serrata*, wh nau, *Elaocarpus dentatus* or tutu, *Coriaria arborea* and then burying the fiber in a "special black mud" with a "rusty colored surface" for one or two days (Mead). It is likely this is iron rich mud and thus an iron tannate dye. Hawaiian natives are said to make a black dye from the soot of burned ripe kukui nuts (Krohn). Soot would not be a substantive dye and no details of the dye process are given, so this may be an iron tannate dye as well.

Other iron-tannate dyes were known to peoples in Africa, Asia, Europe, Aboriginal Australia and ancient Mesopotamia (Brunello, Stuart Robinson, Schetky). Sumac berries without iron mordant are reported to produce a tan or yellow color (Lesch, Van Stralen) and the Ojibwa Indians used sumac, *R. glabra*, without iron as a yellow dye (Brunello). The flavonol fisetin (Fig. 6.12e) occurs in sumac heartwood as a complex with tannin but no mention is made of the berries (Mayer & Cook). Fisetin is most likely the dye chemical responsible for the yellow color obtained with sumac berries as well.

Sumac Iron-Tannate Recipe for Cotton (adapted from Liles)

Bath 1- Sumac Bath or Tannin Bath:
Sumac Bath (Buchanan): Simmer 50 g sumac leaves in 2 L of water for 2-3 hours OR
Tannin Bath (Liles): Dissolve 6 g. (ca. 1 Tbs.) tannin in 2 L hot water.

Bath 2- Washing Soda Bath: Dissolve 7.5 g (ca. 1.5 tsp.) washing soda (sodium carbonate) in a small amount of hot water. When completely dissolved, dilute to 2L with warm water.

Bath 3-Iron Bath: Dissolved 6 g. (ca. 1.5 tsp.) iron (II) sulfate heptahydrate in 2 L water.

Place wet cotton in Bath 1, the tannin bath and let soak*. Remove cotton and wring it out then place it in Bath 2, the washing soda bath, for 10 minutes. Remove and wring then place the wet cotton in Bath 3, the iron bath, for 30 minutes. Repeat all steps until desired darkness of black is obtained.

*For the first cycle, the cotton should be soaked in the tannin bath overnight. For subsequent cycles it should be soaked for 1 hour. After dyeing is completed, wash well with detergent, rinse and dry. This method can be adapted for use with wool by eliminating the washing soda bath (Liles). Wool is damaged by prolonged exposure to alkaline pH (Aspland). Liles cautions that this treatment is hard on the fibers and will shorten their lifespan. It should be used for historical purposes rather than to prepare fibers for use in an art or craft project.

Sumac Berry Yellow Dye Recipe for Wool (adapted from Androsko)

2 L of sumac berries
distilled or soft water

Place berries in large beaker, enamel or stainless steel pot. Add sufficient water to cover the berries and let soak for 1 hour. After soaking, heat the bath to boiling and boil for 30 minutes. Strain the bath with a wire strainer to remove the berries and dilute to a volume of 4 L. Let cool to tepid. Dye alum mordanted wool by adding wetted wool to the tepid bath and slowly raising the bath to a simmer. Simmer 30 min.

Bark Dyes

Barks were used as dyes all over the world. Native Americans used them before the arrival of settlers (Kraemer) and Colonial Americans used them as well (Androsko). Peoples of Hawaii (Krohn) and Africa (Hindmarsh) also used bark dyes. It is likely that every culture experimented with bark dyes to some extent. A generic "bark" recipe is given here because of the multicultural appeal of bark. The dye chemicals in most cases are tannins (Fig. 6.15) that will produce browns and grays, but flavonoids or quinoids may or may not be present depending on the species of tree. Students may enjoy experimenting with different trees to see what colors are produced.

Generic Bark Dye (adapted from Kraemer)

Shred 120 g of bark and soak overnight in a large beaker, enamel or stainless steel pan. with sufficient warm water to cover the bark completely. The next day boil for 1 hour and strain. Add hot water as necessary to bring the volume to 4L. If dyeing wool the solution must be cooled first to avoid felting the wool but may be used hot for cotton. Add wet yarn or fiber and slowly return dyebath to simmer. Simmer for 1 hour. Unmordanted materials may be used or alum mordanted wool or cotton. It is generally not necessary to use alum/tannin mordant on cotton because the bark will contain tannins, however if you have already prepared alum/tannin mordanted fibers they may be used.

Indigo Blue

The plant genus *Indigofera,* commonly known as indigo, is native throughout the world and was a common source of the dye chemical indigotin (Liles, Cannon & Cannon, Stuart Robinson, Brunello). Some of the most common species are *I. tinctoria,* native to India (Liles) and widely grown in the tropics (Cannon & Cannon), *I. suffruticosa,* native to Mexico and South America (Cannon & Cannon, Liles) *I. arrecta,* native to Ethiopia (Cannon & Cannon) and *I. australis,* native to Australia (Cribb & Cribb). There are also many related genera that can be used to produce indigo dye including woad, *Isastis tinctoria* found in Europe and north Africa, *Polygonum species.* found in China, Japan and central Asia (Brunello, Cannon & Cannon, Liles), *Lonchocarpus* native to Africa (Gerber, Jill Goodwin), *Marsdenia* in Sumatra (Gerber, Liles) and *Nerium* in Coromandel (Gerber), India and the Far East (Liles). Plants of still additional genera have been used throughout Asia and the Pacific Islands, North America (Gerber, Brunello), the Nordic regions of Europe (Brunello), and Australia (Gallagher) to produce indigo dye.

While the initial chemical extracted from different plants varies from the indican extracted from *I. tinctorium,* all are chemically related and ultimately produce the soluble indoxyl precursor in the vat fermentation process. Upon exposure to air, this soluble substance is oxidized to the dimer indigotin, the dye chemical in indigo dye. This process is illustrated in steps 1 and 2 of fig 3.1. The insoluble indican precipitates and can be dried for later use, or can be converted to the soluble leuco salt *in situ.* The dyeing process beginning with indigotin is shown in steps 3-5 of fig 3.1. Indigotin is first reversibly reduced to the insoluble leuco compound known as white indigo and then converted to the soluble sodium salt of the leuco compound by raising the pH. The soluble salt can be adsorbed by the fabric. The indigotin responsible for the blue color of indigo dyes is reformed by oxidation in the last step of the process (Christie, Epp, Trotman). The insoluble indigotin is physically trapped in the fabric. Other related plants produce related pigments that undergo similar processes to produce indigotin.

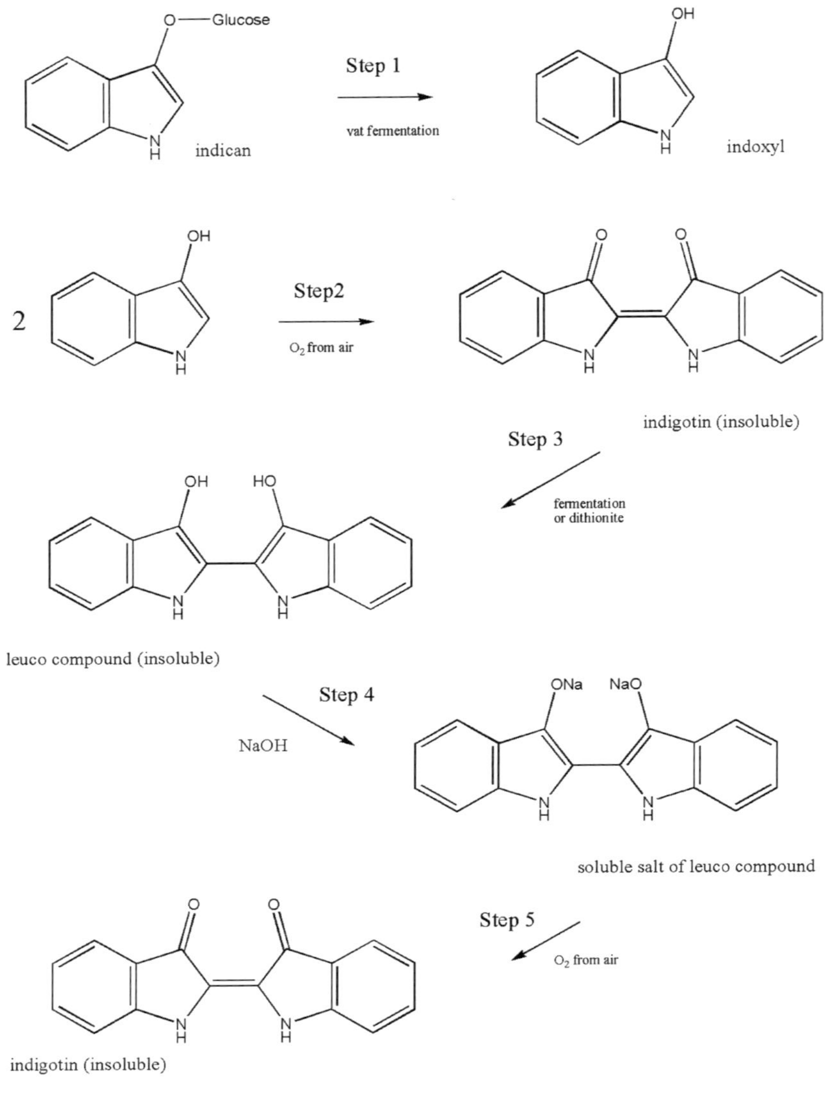

O—Glucose

indican

Step 1

vat fermentation

OH

indoxyl

2

OH

Step2

O_2 from air

O O

indigotin (insoluble)

Step 3

fermentation
or dithionite

OH HO

leuco compound (insoluble)

Step 4

NaOH

ONa NaO

soluble salt of leuco compound

Step 5

O_2 from air

O O

indigotin (insoluble)

Figure 3.1 The extraction of idican, a glycoside of indoxyl and glucose, from *Indigofera sp.* and its conversion to the dye chemical, indigotin are shown in Steps 1-2. Steps 3-5 show the dyeing process where Indigotin is first reduced to the insoluble leuco compound known as white indigo. The leuco compound is converted to the soluble sodium salt which can be adsorbed by the fabric. The indigotin responsible for the blue color of indigo dyes is produced by oxidation in the last step of the process. Other related plants produce pigments related to indican that undergo a similar process to produce indigotin.

Cultures and peoples known to use indigoid dyes from native plants include South and Central American Indians, North American Indians (Brunello) including Mexican Indians (Harris, Robison, Sayer), most of Asia (Harris, Brunello) including China, Japan (Cannon & Cannon, Gallagher, Gerber, Toguchi) and Laos (Cheeseman), the Pacific Island regions including Timor, Sumatra, Burma, Borneo, New Guinea and Melanesia (Gerber, Jill Goodwin, Brunello, Harris, Stuart Robinson), much of North, Central and Western Africa including Egypt (Cannon & Cannon, Gerber, Jill Goodwin, Stuart Robinson) and Nigeria (Isaacs), and the early Europeans (Brunello, Cannon & Cannon). Still more cultures used indigo obtained by trade including the Navajo (Bryan & Young) and Hopi (Colton) of the American southwest, who traded with Mexico (Colton) and the early American settlers (Bronson) who purchased indigo from European suppliers with links to the east (Bronson). While some contend that indigo was not used in pre-Colombian America (Brunello) most dispute this and give evidence of ancient use of indigo in the Americas (Bryan & Young, Colton, Harris, Stuart Robinson). There are reports of "wild indigo" being used by emancipated African-American slaves in the American south (Waldvogel '94) and by the Arcadian or "Cajun" settlers of Lousiana (Gerber). Wild indigo, *Baptisa tintoria*, among others, does grow in North America but the genus and species of the plants used by African-Americans and the Cajuns is not specified.

While it is possible to grow indigo producing plants, finding the seeds can be difficult and the plants are not always successful. In most cases the plants need to be used fresh (Buchanan) or preserved by a time consuming process known as "balling" (Cannon & Cannon, Jill Goodwin). An alternate process for preservation was sometimes used in which the indigo is extracted by steeping plants, and then oxidized and precipitated (Gerber). This is presumably accomplished by stopping after step 2 of fig 3.1. The indigo in this stage is of higher purity than the balling method and can be stored more easily (Gerber). Either way, extracting the dye stuff from the plant is a long, labor intensive process. Fortunately, it is quite a bit simpler to purchase dried natural indigo or synthetic indigo (indigotin) and use it to dye cloth. The dye chemical is identical to the pure form obtained from the plant and the chemical process is similar to that our ancestors would have used. Most of them purchased or traded for indigo and did not start with the plant itself. Recipes using the fresh leaves are given elsewhere (Buchanan, Cannon & Cannon).

Traditional recipes use stale urine and a vat fermentation process to produce ammonia and ultimately a chemically reductive environment. Both warm vat (50-60°C) and cold vat (ambient temperature) methods have been used with success (Thompson & Thompson). Liles gives several versions of the traditional vat recipe for readers interested in a historically accurate recipe including one version using simulated urine (Liles) rather than the authentic item. The recipe given here, like most indigo recipes found in dye books, uses the chemical reducing agent known as sodium dithionite (sodium hydrosulfite) and sold by dye suppliers as Spectralite™. The results using dithionite are faster and more predictable than the traditional alkaline fermentation methods and avoid the unpleasant task of collecting, storing and handling stale urine. Sodium dithionite decomposes upon exposure to air so larger amounts may be required if using a previously opened bottle.

Easy Recipe for Dyeing with Synthetic* Indigo:

Make an indigo suspension by dissolving 0.4 g of synthetic indigo in 100 mL of distilled water in a 2L beaker. Make a 10% sodium hydroxide solution by dissolving 30 g of sodium hydroxide in 270 mL of water. Add 250 mL of the sodium hydroxide solution to the indigo suspension and heat to 60°C. Maintain the solution at 60°C for an additional 30 min. then turn off heat and allow solution to cool to 45°C. Add 20 g of sodium hydrosulfite (aka sodium dithionite, Na2S2O4) and stir until dissolved. Dilute the solution to total volume of 650 mL. The solution will initially be greenish but will fade to yellow. The blue dye will be produced on the fabric when it is oxidized by exposure to air upon removal from dye vat. This dye should be prepared and used on the same day.

Dye the fibers by placing wet fibers in the indigo bath for 1 hour. It is not necessary to heat the bath, although raising the temperature to about 60°C may help with cotton. In the case of wool, it is safer not to heat it as the alkaline environment and high temperature tend to damage the wool fibers. Remove the fibers from the dyebath but do not rinse. Let them air oxidize to blue. Repeated dips in the dye may be needed to attain a darker blue, or the fiber may be allowed to stand in the vat for as long as a day or two. The first rinsing of the indigo dyed item should be a minimum 12 hours after removal from the dyebath and longer if practical (Thompson & Thompson).

* This recipe is designed for synthetic indigo however, natural indigo can be used if allowances are made for reduced purity of the natural product. Gerber reports natural indigo preserved by the extraction and precipitation method to be about 48% pure (Gerber). Using this as a guideline, then 0.8 g of natural indigo would be required in the above recipe. I have not tested the procedure with natural indigo and under the best circumstances, the purity of the natural product will vary giving varying results.

Quercetin and Quercetagetin Yellows and Oranges

Quercitron is a dye stuff extracted from the bark of Oak species, particularly black oak, *Quercus velutina.* It contains dye chemicals quercetin (Fig 6.12a) and quercetagetin (Fig 6.12b) (Cannon & Cannon). Both are chemicals known as flavonols. These two dye chemicals are also found in a number of dye plants throughout the world (Mayer & Cook, Swain 1980) and used as dyestuffs by a number of different peoples. Because the same two chemical compounds are the active ingredients in many dyes, the discussion of them will be combined here.

The dye mixture quercitron is found in the bark, leaves and wood of oak, *Quercus sp.*, and was used throughout America from the Aztecs through modern times (Bronson, Bryan & Young, Cannon & Cannon, Findley, Liles, Ronsheim, Schetky, and Waldvogel '90). It was the basis of the colonial American dye industry and was used commercially for centuries (Butterworth, Findley, Liles). It was exported to Europe as a replacement for weld, *Reseda luteola* (Cannon & Cannon). There are references to the much earlier use of oak bark dye in Europe by Descordes and Pliny (Schetsky) and it is cited as a natural dye used in Japan (Schetsky) although she does not say whether it was important historically.

Quercetin is also one of the major dye chemical found in onion skins (*Allium cepa*) along with several related flavonols (Cannon & Cannon). Onion skins are used as a dye stuff in Africa (Brunello), various European countries (Schetky) the southeast U.S. (Liles, Schetky) and by many of our American ancestors including my grandmother. The "Navajo Dye Chart" art poster lists wild onion, and red and brown onion skins as dyes used by the Navajo (Myers) although none of these is listed by Bryan & Young as traditional Navajo dyes.

Quercetin is also found in Goldenrod (*Solidago canadensis* and *S. sp.*) along with the related flavonol, kaempferol (fig 6.12d) (Cannon & Cannon). Goldenrod was known as a dye plant to early American settlers (Cannon & Cannon, Liles, Ronsheim). Quercetin is also found in *Euphorbia sp.* used in Turkey to dye "Oriental" carpets (Gallagher). It is still used by the Navajo (Bryan & Young, Cannon & Cannon, Myers), Hopi (Colton) and natural dye hobbyists (Buchanan, Liles). Quercetin and related flavonols are also found in large quantities in instant tea powder (Cannon & Cannon) and presumably in tea leaf as well. It probably accounts for the orange-browns obtained with tea dyes. Quercetin is even found in the stalks of tomato plants (Mayer & Cook).

Quercetin has been found in at least two species of Eucalyptus leaves, *E. nichloii* (Glasson & Glasson) and *E. microrhyncha* (Cribb & Cribb). Many other eucalyptus species have been shown to contain other flavonols including some quercetin derivatives (Cannon & Cannon). All eucalypts reported are said to dye a range of colors from buffs, yellow, orange and reds with buffs and yellows being most common (Carman, Glasson & Glasson, Hallett, Kraemer, Lloyd, Martin, Milner, and Neilsen). This is in keeping with flavonol dyes in general and quercetin in particular.

The second oak dye chemical, quercetagetin, a derivative of quercetin, is also found in Marigold flowers (Cannon & Cannon, Mayer & Cook) along with a related flavonol patuletin (Fig. 6.12f) (Cannon & Cannon). The plant called "French Marigold," *Tagetes patula*, is actually native to Mexico and many other *Tagetes sp.* are found in South America (Cannon & Cannon). One marigold species, *Tagetes erecta*, is native to Africa (Cannon & Cannon) and a related plant called Bitterball, *Tagetes micrantha*, is found in the American southwest (Bryan & Young, Cannon & Cannon). Unlike quercetin, quercetagetin is fairly limited in its distribution. In flowers it appears to be restricted to the composites (Harborne 1976). It is still included in the multicultural section because there is evidence that marigold flowers and similar composites were used as dye plants by many cultures, as detailed below.

Bitterball is known as a yellow dye to the Navajo but is not a popular dye because the plant is a sacred medicine plant to the Navajo (Bryan & Young). Marigold species are often used by modern hobbyists and produce an easy fast dye (Buchanan, Casselman, Lesch, Liles, Van Stralen). There is no specific evidence that the ancient Africans used Marigold as a dye or that the pre-Colombian Native-Americans used marigold or goldenrod. However garments dyed with unknown yellow plant dyes have been found in pre-Colombian sites in the Americas (Colton, Harris) and a wide range of plant colors was used by the Ewe and Asante peoples of Ghana (Harris). Given the wide color palette of dyes known to these peoples (Colton, Harris) and their sophistication in the art of dyeing, it seems highly probable that they were well aware of the dye properties of these plants.

Recipes for Marigold, Onion and Eucalyptus are given here and recipes for Goldenrod dye and Oak dye are found in Chapter 4. All five dyestuffs contain quercetin, quercetagetin or related flavonols however recipes using the North American species are presented in the North American Dye palette, rather than the Multicultural Dye selection of this chapter. Although quercetin is the major dye chemical in all three plants presented below, other pigments are present and the color of the fiber differs in all three cases. Samples dyed with marigold, onion and eucalyptus are shown in photos 2a-d in the color insert section of this book.

Easy Marigold Dye

Marigolds contain a natural pesticide so the process should be carried out in a chemical fume hood or outside with adequate ventilation! Place about 500 mL of dry or 1 L of fresh marigold flower heads in a 6-8 L beaker or a stainless steel or enamel pan and add sufficient distilled water to completely cover them. Heat to nearly boiling and simmer for 30 minutes. Strain dye bath to remove the flower heads and reserve the liquid. Add water as needed to make a 4 L dyebath. Place alum mordanted wool or alum/tannin mordanted cotton into the dye bath and heat to near boiling. Continue to simmer for 15-60 min. until desired color is reached. Different colors may be obtained from fresh and dry plants and depending on whether or not the green sepals and stems are removed from the flowers. They will range from yellow-orange to a greenish yellow.

Even Easier Onion Dye:

Place about 1 L of loose dry brown onion skins* in a 6-8 L beaker or a large stainless steel or enamel pan. Add sufficient distilled water to easily cover the skins. Bring to a boil and simmer 30-60 min. stirring occasionally. Strain with wire mesh to remove the onion skins. If desired, the skins can be left in the dyebath during the dyeing process and will produce a mottled look on the fabric. Alternately, the onion skins can be placed in an old "knee high" nylon stocking prior to heating. This makes removal easier but does impede water getting to the skins and may require more stirring and longer heating. Add water as needed to make a 4 L dye bath. Add wet fiber to the dyebath and heat to nearly boiling. Simmer 30-60 min. This dye will work on unmordanted fibers, but will produce a deeper orange color on alum mordanted wool or alum/tannin mordanted cotton. Different varieties of onion skin will produce slightly different colors caused by the various anthocyanins present in the onion skin. For a further discussion of anthocyanins as dyes, see Ch. 5 and 6.

*A free or inexpensive source of dry onion skins is the grocery store. Approach the produce manager and volunteer to clean out the onion sales tray in exchange for free skins. If you are reluctant to do this, fill a plastic produce bag with dry skins and pay for them at the onion price. They don't weigh very much so this is a very inexpensive purchase.

Silver Dollar Eucalyptus Dye* -Method I Yellow Dye (adapted from Kraemer)

Break up 120 g of fresh silver dollar eucalyptus leaves into a 4 L beaker or stainless steel or enamel pot and add 2 L of water. Let stand overnight. The next day bring to nearly boiling and simmer for 1 hr. Strain to remove the leaves. Add alum mordanted wool and alum/tannin mordanted cotton and simmer for 1 hr. Kraemer says the dye will produce a yellow color or yellow orange color.

* The silver dollar eucalyptus (*Eucalyptus cordata*) is specified because it can be easily purchased from most florists. Hundreds of species of eucalyptus yield dyes and if you have access to other species, by all means try them with the General Eucalyptus recipe below.

Silver Dollar Eucalyptus Dye -Method II Red Dye (adapted from Milner)

Follow the recipe above with the following changes. Omit the soaking overnight while increasing the simmering time to 2 hrs. Following this method is reported to yield a red dye. Many authors do report the same eucalyptus species producing red and yellow dyes depending on the length of simmer and the amount of fiber used (Carman, Glasson & Glasson) but Carman (1978) has also observed very different colors from time to time using the same species following the same recipe. Her experiments conclude that whether the leaves are fresh or dried, the weather, season, amount of rainfall and geographic area where the plants were grown are all factors in determining the final dye color from the same species even when the dyeing is conducted by the same person. If the silver dollar eucalyptus leaves are purchased, none of these factors will be known and results can be expected to vary from time to time. Indeed a wide range of colors from dried sliver dollar eucalyptus leaves have been found in New Zealand (Neilsen).

Generic Eucalyptus Leaf Dye (adapted from Carman 1978 and Glasson & Glasson)

This recipe can dye about 50 g of fabric if a yellow dye is produced*. Break up 100g of dried Eucalyptus leaves in a 4 L beaker or large stainless steel or enamel pan. Add sufficient water to completely cover to a depth of several inches and let stand overnight, turning over the leaves a few times. Cover the pan and very slowly bring near to a boil and simmer for ½ hr. to 1½ hrs stirring occasionally*. Remove from heat and strain with a wire mesh to remove the leaves. Added wet alum mordanted wool or cotton to the dyebath and heat nearly to boiling. Eucalypts produce a substantive dye due to tannins in the leaves so it is not necessary to use tannin mordanted cotton, however if you have prepared alum/tannin mordanted cotton, it may be used. Simmer 20 min. for wool or 30 min. for cotton. Let the fabric stand in the bath until it cools slightly, then remove, rinse and dry.

* Carman (1978) says that for yellow dyes, ½ hr. is generally sufficient simmering time. If the dyebath begins to turn orange continue simmering for another 10 min. If it appears to be turning red, continue simmering for the full time.

I have produced "old gold" to buff dyes in my limited attempts with eucalypt dyes. According to Carman, if you are lucky enough to produce a red dye with the species of eucalyptus used, the dyebath can be used again to produce a second batch of orange fiber and a third batch of yellow fiber. In general considerably less fiber can be dyed red or orange. Carman (1978) recommends 4 -10 fold weight ratio of leaves to yarn for orange dyes and 10-16 fold ratio of leaves to yarn for red dyes rather than the 2 fold ratio given here in the general recipe. Several references have species specific information about dyeing with eucalyptus and may be consulted if you have several species available (Cannon & Cannon, Carman, Glasson & Glasson, Lloyd).

<u>Barberry Yellows</u>

Many plants in the barberry family produce the dye berberine (fig. 6.19). The tree *Mahonia japonica* is native to China but cultivated in Japan as well. The wood and the bark are sources of yellow dye (Cannon & Cannon). Oregon grape (*M. aquifolium*) is a native of the Pacific northwest region of America and also contains this dye (Cannon & Cannon). The early Egyptians (Stuart Robinson), ancient Europeans (Schetky) and many American Indians (Brunello) used Barberry roots as a dye stuff. The Nez Perce of the American northwest used the roots of Oregon grape as a dye (Harris) while the Navajo use the entire plant to produce a greenish-yellow dye (Bryan and Young). It is listed as a traditional dye plant in Greece (Schetky).

Barberry Wood Yellow Dye Recipe for Wool (adapted from Wiegle)

To dye 30 g of wool (about 1 oz.), place 2 cups of chopped barberry stems or roots and 1.5 L of distilled water in a 4 L beaker or large stainless steel or enamel pan. Soak for 1-3 days then bring the pot to a boil and simmer for 2 hr adding water as necessary to maintain the volume. Strain the dyebath to remove the wood and cool. If necessary add water to make 1L of dye. Add wet alum mordant wool or alum/tannic acid mordanted cotton to the dyebath and simmer for 30-60 min.

Barberry Leaf Yellow Dye Recipe for Wool (adapted from Cannon & Cannon)

Place 2 cups of chopped leaves in an open container with 1.5 L of water and let stand 1-2 weeks for fermentation to occur. Add wet alum mordanted wool or alum/tannin mordanted cotton and raise the temperature to 50°C then remove from heat and let stand 48 hrs.

Chapter 4
North American Native Dye Plants and Dyes

Several important dyes are found in plants native to North America. Some of these dyes are found in smaller amounts in related species from other parts of the world. They are listed together as American dyes either because the species containing the highest concentration of dye is an American species or because the genus is more common in North America than in other parts of the world. This section includes five dye plants still in common use today by natural dye enthusiasts. Each is easy to collect or grow and produces a reasonably fast dye.

Black Oak

Black oak and other oak species contain quercitron (quercetin and quercetagetin) and have been used as dye plants in America from at least Colonial days until recent times (Cannon & Cannon, Liles, Findley, Bronson, Bryan & Young, Ronsheim, Waldvogel '90). See a more complete discussion of the use of oak and related dyes under "Quercetin and Quercetagetin Dyes" in the Multicultural Dye Palette in Ch 3.

Black Oak Yellow Dye Recipe (adapted from methods of Kraemer and Liles)

Chop about 50 g of bark from small branches and soak in 4 L of lukewarm water overnight. Alternately the soaking bath can be raised to near boiling then covered and removed from heat to soak overnight. The next day return to heat and simmer 1-2 hrs. Remove the bark by straining and add water as needed to maintain the 4 L volume. Add alum mordanted wool or alum/tannin mordanted cotton and simmer 30 – 60 minutes. Cotton will require less time than wool.

Black Walnut

Black walnut, *Juglans nigra,* is an American species (Cannon & Cannon) but other walnut species grow in Europe and mention of walnuts as a dye stuff goes back to Pliny and

Descordes in Europe (Cannon & Cannon, Schetky). Wild walnut, *J. major*, is used as a dye by the Navajo Indians (Bryan & Young). Black walnut and butternut, *J. cinerea*, were used by Colonial Americans (Bronson, Cannon & Cannon, Liles) and until recently in rural America (Findley, Ramsey and Waldvogel,Ronsheim, Valentine, Waldvogel '90, Waldvogel '94). Great Lakes Indian tribes of Canada used walnut as a traditional dye (Schetky). The dye chemical juglone (fig 6.14a) is found in all walnut species and in butternuts (Cannon & Cannon, Mayer & Cook) but is most concentrated in black walnut. All parts of the tree contain the chemical (Cannon & Cannon, Leistner) and can be used as a dye source but the juglone is most concentrated in the nuthulls (Cannon & Cannon). This is still a popular dye with hobbyists because it is easy and fast due to the tannins present in the walnut materials (Androsko, Casselman, Lesch, Van Stralen). Black walnut dye is often used to dye basketry materials and other wooden folk art items. Fabric, fiber and a basket dyed with black walnut are shown in photographs 3a-c of the color insert section of this book

Black Walnut Brown Dye Recipe for Fabrics

Most recipes call for removing the walnut hulls from the nut, however this is a difficult and messy process. It is faster and easier to use the nuts intact, either fresh or frozen and thawed before use. Take 10-20 walnuts and slice or pierce the hulls in many places to make it easier to extract the dye from the husks. It's a good idea to wear gloves when collecting and piercing the hulls to avoid staining your fingers. Place the prepared hulls into a 6-8 L beaker or an enamel or stainless steel pan and add sufficient water to cover them. Bring to a boil and simmer 30 minutes. Strain to remove the walnuts and add more water if necessary to make a 4 L volume for dyeing. Add wet cotton or wool to the dye bath and return to a boil. Simmer 30 minutes. Walnut is a substantive dye, meaning no mordant is necessary but alum mordanted wool and alum/tannin mordanted cotton may be dyed to a deeper and slightly different color of brown.

Black Walnut Dye Recipe for Basketry and Related Items

Several years ago a basketry teacher at the Cincinnati Nature Center used this process to dye ash baskets. The Center offered several classes a year and so walnuts were collected as they fell, put into a large container and covered with water. The dyebath was large but shallow, about the size of the small molded plastic childrens' swimming pool. The bath sat and fermented during the year and each fall more walnuts were added. Most dyers will not be dyeing as many baskets as a class would, so the procedure is adapted here for occasional use.

Prepare the walnut dye as described in the recipe for fabric above. After removing the walnuts let the dyebath cool. Transfer the dye to a container large enough to allow one side of the basket to be immersed in the dye bath. It's a good idea to wear rubber gloves during the dipping process. It is not necessary to immerse the entire basket at one time. Instead turn the basket to each side or remove and re-dip as necessary to dye all four sides and bottom of the basket. Hold the basket by the side or bottom to dip the handle. The wood or dried reed materials of the basket will take up the dye much more quickly than fabric.

*This is traditionally done immediately after making a basket. When making a basket, the wood slats are soaked several hours to make them more pliable and usually the basket is still damp when dyed. It may be necessary to soak a purchased basket in warm water for a short time before dyeing to simulate the basket making process and ensure more even dying of the basket. Whether or not the basket needs to be soaked and for how long will depend on the basket material. Some experimentation may be necessary to achieve the desired results.

Black Walnut Black Dye Recipe

Follow the same procedure as described for walnut brown dye but use iron mordanted wool or cotton. Alternately use alum mordanted wool or alum/tannin mordanted cotton and add the iron to the dyebath where it will act as a simultaneous mordant. The second method is helpful if alum-mordanted fabric has been prepared for use with the other dyes. Be sure to measure the iron as carefully as for a separate iron mordant. Too much iron it will "tender" or damage the fabric shortening its lifetime. The holes in the black fabric of antique quilts are often caused by tendering of the fabric caused by iron mordants. This recipe will produce a surprisingly deep black on cotton and wool.

Coreopsis or Tickseed

The annual wild flower Tickseed or Coreopsis (*Coreopsis tinctoria*) and related coreopsis species are found mainly in North America (Cannon & Cannon). Related plants *Cosmos suphureus* and *Dahlia pinnata* are native to Mexico and Central America and exhibit similar dye properties (Cannon & Cannon, Buchanan). There is good evidence that both were used as dye plants in Mexico prior to the arrival of Europeans (Cannon & Cannon, Schetky). *C. tinctoria* can be grown from seed but the seeds are not always easy to find. These annuals reportedly can self-sow (Buchanan) but I have not found this to be a reliable method and recommend planting them each year or planting large crops as needed and drying the flowers for storage.

Other coreopsis perennial species commonly used as garden plants can also be used for dyeing (Buchanan, Cannon & Cannon) and produce similar colors although more flowers may be required to achieve the same depth of color. Coreopsis are easy to grow from seed and produce abundant flowers. The perennials will return year after year with little maintenance. The important dye chemicals are a class known as anthochlors that encompass two structural types, chalcones and aurones. Large amounts of both chalcones and aurones are found in the coreopsis subtribe (Boehm). Chalcones are found in a variety of families (Harborne 1966) however aurones are never found in lower plants (Harborne 1966) and appear to be rather rare (Swain 1976).

The dye chemicals in *Coreopsis* and related plants include glycosides of the chalcones butein (Fig 6.17b), marein (Fig. 6.17c) and stillopsin (Fig 6.17d) and glycosides of the aurones sulphuretin (Fig. 6.17f), marimetin (Fig. 6.17g) and leptosin (Fig. 6.17h) (Boehm, Cannon & Cannon). They also contain anthocyanins (Cannon & Cannon). *C. tinctoria* have red coloration of the inner petal and yellow as the background and anthochlors and anthocyanins are often found together in patterned petals (Harborne 1976). Carotenoids (fig 6.20) are also present in the petals of coreopsis species (Harborne 1976). Cosmos also contain anthochlors (Boehm) and probably carotenoids as they are also commonly found with anthochlors (Harborne 1976). Dahlias contain aurones and chalcones as the only coloring agents which is unusual (Boehm, Trevor Robinson, Harborne 1976). The anthochlor pigments in Dahlia and Cosmos account for the similar dyeing properties they share with coreopsis species. Because the major dyes are anthochlors, the dye colors will be more red at alkaline pH (see Ch. 8).

Coreopsis Yellow or Orange Dye Recipe:

Gather coreopsis flowers daily by pinching them off at the flower base and use them fresh or dry them for later use. They can be dried easily and quickly in a few days on a wire rack or in a few hours with a food dehydrator. When ready to dye, place about 1 L of fresh or ½ L of dried flowers into a 8 L beaker, stainless steel or enamel pan (more flowers will be helpful if using a species other than C. tinctoria). Cover with 4 L of water and bring nearly to a boil. Simmer for 30 minutes then strain to remove plant materials. Add water as needed to retain a 4 L volume. Add wet alum mordanted wool or wet alum/tannin mordanted cotton and simmer 15 to 60 min until desired color is obtained. Unmordanted fibers may also be used but produce a yellower color. Both Buchanan and Cannon & Cannon report that redder hues can be obtained on fibers by making the dye bath slightly alkaline or by rinsing the fibers in an afterbath of sodium bicarbonate. The chemistry explaining this color change is detailed in Ch. 8.

Goldenrod

Goldenrod, *Solidago sp.*, was known as a dye by early American settlers (Cannon & Cannon) and was used until recently in Ohio (Ronsheim). It is still used by the Navajo (Cannon & Cannon, Bryan & Young, Myers), Hopi (Colton) and modern hobbyists (Buchanan, Cassellman, Lesch, Liles, Van Stralen). See the complete discussion of goldenrod dyes under "Quercetin and Quercetagetin Dyes" in the Multicultural Dye Section, Ch. 3.

Goldenrod Yellow Dye Recipe (modifed from Colton)

Place 90 g of goldenrod blossoms into a 10-12 L beaker or large stainless steel or enamel pan. Add 6 L of water and bring to nearly boiling. Boil for 1-2 hr then strain to remove the flowers. The dyebath should be about 4.5 L in volume. Add alum/tannin mordanted* cotton or alum mordanted wool and simmer gently 1-2 hrs. to yield a fast yellow dye.

* This is modified from the original Hopi recipe which does not call for premordanted cotton but instead the addition of 3 "double handfuls" of native ground alum to the dyebath just before adding the cotton fibers (Colton). This suggests that simultaneous mordanting of the cotton will work and that extermal tannins are not needed to produce a fast dye on cotton. The major difficulty, of course, is in estimating how much pure alum is equivalent to the native alum in the original recipe. Colton reports that one double handful is about 100 mL but the purity of the native alum is not reported (Colton). Interestingly, the Hopi only dye cotton not wool with this dyestuff. Recipes from other sources use pre-mordanted fibers

and most use wool rather than cotton. The more typical wool and cotton mordants are recommended here for convenience, however this recipe could be used in the traditional Hopi manner for cultural accuracy if desired. In that case purchasing raw alum from an earth science or geology school supplier and using about 300 ml (about 1 1/3 cups) of the ground alum would be an appropriate starting point.

Osage Orange

The Osage Orange tree, *Machura pomifera*, is native to south-central U.S. (Cannon & Cannon, Liles, Ohio DNR). The wood is usually used for dyeing and all the wood parts including the roots contain the dye stuff (Cannon & Cannon, Liles). The most important dye chemical in the wood is morin (fig 6.12c) (Cannon & Cannon) and in the fruit are osajin (fig 6.18a) and pomiferin (fig 6.18b) (Cannon & Cannon, Harborne 1976, Mayer & Cook). Morin, the dye chemical found in the wood is the same as that extracted from the S. American "Dyers' Mulberry," *Chlorophora tinctoria*, (Cannon & Cannon, Liles, Mayer & Cook). The dye extracted from *C. tinctoria* is usually called "old fustic" and contains more of the dye chemical than *M. pomifera* does (Cannon & Cannon). Small wood chips or sawdust work best for use as a dye stuff (Liles) but small branches can also be used to produce similar but less orange colors (Cannon & Cannon). The dye was used for the khaki uniforms of World War I (Schetsky) and is still used by hobbyists for a good yellow (Liles, Van Stralen). Although the fruits are not traditionally used for dyeing, they are much easier to collect and use than the wood so recipes for both are given here.

Osage Orange Wood Dye Recipe (adapted from Cannon & Cannon and Liles)

Chop 100g of small branches and soak overnight in 4 L of water or add 50-60 g of wood chips or sawdust to 4 L of water in an 8 L beaker or enamel or stainless steel pan. Boil for 2-3 hours then strain to remove the plant materials. Add water as needed to maintain a volume of 4 L and cool to 55 - 70 °C. Add wet alum mordanted wool or wet alum/tannin mordanted cotton and soak for 30 min. while maintaining the temperature between 55-70°C.

Osage Orange Fruit Dye Recipe

Take 10 osage orange fruits and slice or pierce the hulls in many places to make it easier to extract the dye. Place them into a 12 L beaker or a large enamel or stainless steel pan and add sufficient water to cover them. Soak overnight. The next day bring to a boil and simmer 30-60 minutes. Strain to remove the fruits and add more water if necessary to make a 4 L volume for dyeing. Add wet alum mordanted wool or wet alum/tannin mordanted cotton and simmer for 30 -60 min.

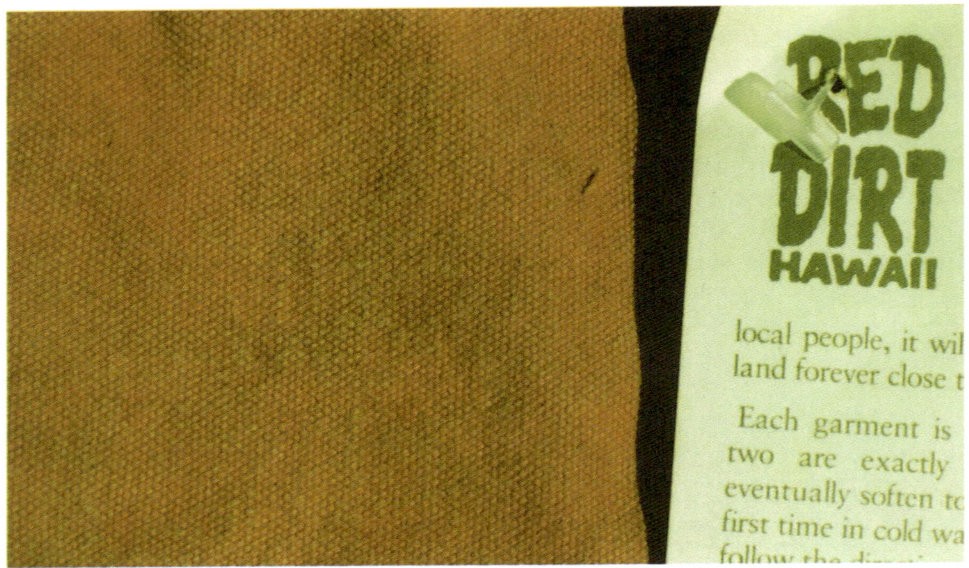

Photograph 1: Paradise Sports "Red Dirt Bag" is colored with an iron buff dye

Photograph 2a: Cotton fabric dyed with a quercitin containing dye from marigold flowers

Photograph 2b: Marigold dyed wool (left) and cotton (right) yarns

Photograph 2c: Cotton fabric dyed with a quercitin containing dye from yellow onion skins

Photograph 2d: Eucalyptus leaves contain quercitin and were used to dye a cotton fabric sample and to tie-dye a cotton t-shirt.

Photograph 3a: Cotton fabric dyed with juglone from black walnut husks

Photograph 3b: Ash basket dyed with a cold black walnut dye bath

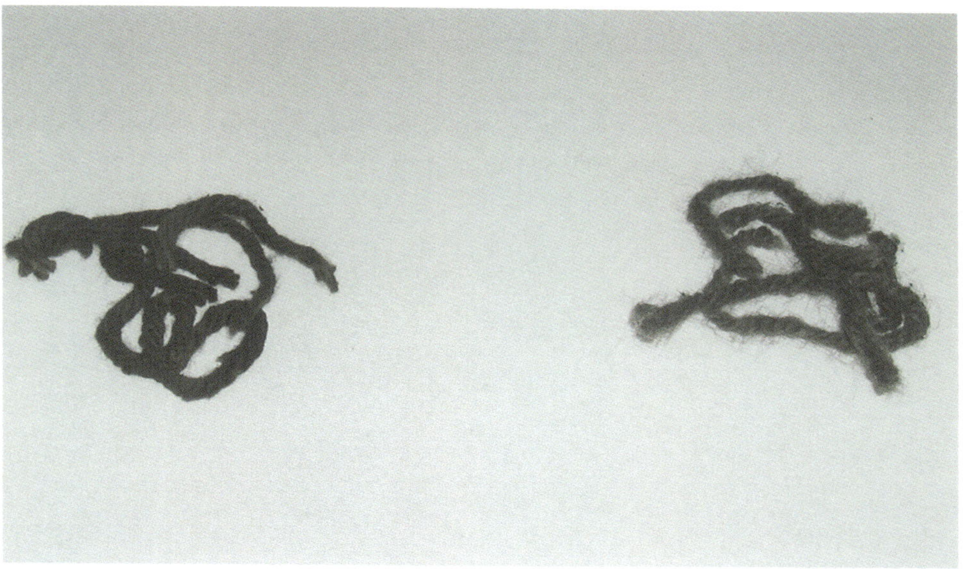

Photograph 3c: A black dye can be produced from black walnuts and iron mordant. Cotton (left) and wool (right) iron mordanted yarns were dyed with black walnut

Photograph 4a: Anthocyanin color in solution varies with pH and mordants. From L to R, rosehip tea with vinegar, with baking soda, with alum, with alum and baking soda.

Photograph 4b: Freshly dyed cotton fabric samples dyed with the anthocyanin solutions shown in figure 4a. Top row L to R: with vinegar, with baking soda. Middle: undyed fabric. Bottom row L to R: with alum, with alum and baking soda.

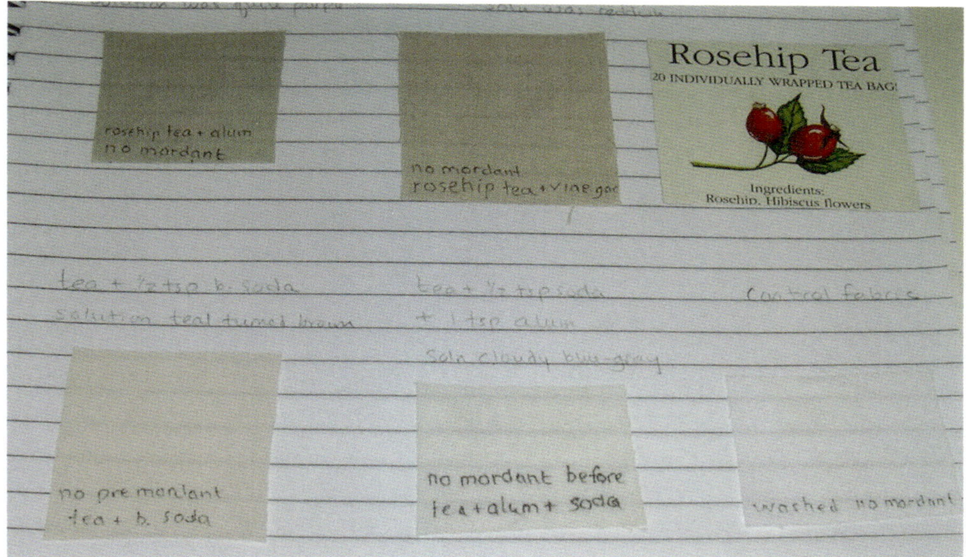

Photograph 4c: Anthocyanin dyes are not fast. The colors fade over time.

Photograph 5a: Cotton fabric dyed with chamomile

Photograph 5b: Cotton fabric dyed with weld

Photograph 6a: Cotton fabric dyed with mint plants. The dye chemicals are most likely tannins.

Photograph 6b: Cotton fabric dyed with tea bags. Tea contains tannins, quercitin and other compounds as described in the text.

Photograph 6c: Quilts and linens can be artificially antiqued using tea dye.

Chapter 5
Natural Dyes Used By Hobbyists and Our Foremothers

Modern hobbyists frequently use several of the dyes discussed in previous sections including black walnut, coreopsis and goldenrod dyes in the North American Section, and marigold and onion dyes in the Multicultural Section. Many of our frugal grandmothers and great grandmothers used natural food dyes, such as onion skins to dye eggs for Easter and coffee, tea or walnut to dye stained household items giving them a new life. A few more dyes used by modern hobbyists or our grandmothers are presented here because they are easy to grow or find and make easy dyes for use with children or students. Most are available throughout the US and Canada. Various anthocyanin dyes and weld were also used by other cultures as well and could be included in a multicultural presentation.

Anthocyanins

Most red, pink and blue flowers, fruits and berries contain anthocyanins (Fig. 6.13) (Cannon & Cannon, Cutright et. al., Harborne 1976, and Mayer & Cook). They are beautiful and often dye fabric wonderful colors but most are notoriously fugitive. The original colors will not be wash-fast particularly with alkaline soaps (Mayer & Cook). They are reported to be light-fast, but a faded-looking pink tinged beige generally appears over a long time. This is probably because of dye breakdown on standing (Mayer & Cook). Keeping their inherently fugitive nature in mind though, anthocyanin dyes can be fun to use. Some have historical importance, such as Pokeberries, Mulberries and Hollyhocks. Pokeweed, *Phytolacca Americana,* grows as a wildflower throughout North America (Cannon & Cannon). Its fruit, the Pokeberry was used by rural African-Americans and European Americans (Findley, Liles, Ramsey & Waldvogel). The seeds and purple parts of the stem contain additional compounds that are very toxic (Cannon & Cannon) so these plants should NEVER be used with children. The dye chemical is cyanidin, an anthocyanin.

Anthocyanins also have a multicultural componenent. Mulberry was used by the Aztecs in pre-Colombian times (Schetky) and Hollyhocks were used in ancient Egypt to produce blue-violets and reds (Trevor Robinson). The Hawaiians use akala, native raspberries, to make lavender and pink dyes and the berries of ‘uki ‘uki, the dianella lily, for blue dyes (Krohn). The Alaskan Tlingit people dyed baskets with berries (Wright) and the magenta dye extracted from the Yeheb leaves by Kenyan peoples (Hindmarsh) may be an anthocyanin. Leaves do contain anthocyanin (Harborne 1976) although some brightly colored flavonoids are reported as well (Goodwin & Mercer).

Anthocyanin dyes are included in this section because, in spite of their fugitive nature, they are still used today by hobbyists attracted by their beautiful if fleeting colors (Androsko, Buchanan, Casselman, Cannon & Cannon, Lesch). Many anthocyanin containing plants have been used for dyes including Blackberries (Cannon & Cannon, Casselman, Lesch, Schetky), Blueberries (Casselman), Cranberries (Casselman, Hess), Elderberries (Cannon & Cannon, Casselman, Lesch, Schetky), Huckleberry (Casselman), Mulberries (Schetky, Lesch), Purple Basil (Buchanan) Hibiscus (Buchanan), Raspberry (Casselman) and Rose (Casselman). There are many convenient sources of anthocyanins including red fruit juices, such as cranberry juice, and herbal tea bags that contain red berries, such as raspberry, and red flowers such as Hibiscus and rose hips. The food sources are much safer alternatives for use with children than Pokeberries. A purple gray color can be attained with fresh purple basil (Buchanan). This dye holds color better than most other anthocyanin dyes I have tried. Fabric dyed with purple basil can be seen peaking out of the walnut-dyed basket in color photograph 3b.

All anthocyanin colors are pH dependent as can be easily demonstrated by adding vinegar or ammonia to the dye bath. The much circulated recipe for using red cabbage as an acid base indicator is based on this pH dependence of the anothocyanins. Red cabbage, *Brassica oleraca,* is among the most highly pigmented plants (Harborne 1976). In general the red colors require acids such as vinegar to be added to the dye bath. Tan, blue or blue-gray colors result if the dye bath is alkaline. The colors can also be modified after dyeing by an acid or alkaline afterbath. Traditionally vinegar is used for an acid bath and washing soda for an alkaline bath, however baking soda or ammonia can also be used. This pH effect can be the basis of interesting experiments for students (see part 4). In the plants however, pH plays a relatively small role in determining flower or fruit color. Other factors including metal complexation and co-pigmentation with flavonoids are more important (Goodwin & Mercer). Samples of fiber dyed with anthocyanins, the pH dependent color changes, and fading of anthocyanin dyes are shown in photographs 4a-c in the color insert section of the book.

Mulberry Dye

Mulberries contain the anthocyanin cyanidin (Harborne 1976) and are reported to produce a violet color on tannin mordanted cotton (Mayer & Cook) although in practice the color may be paler and more tan. Collect 1-2 L of ripe mulberries and put into an 8 L beaker or a stainless steel or enamel pan and add sufficient water to cover them. Rubber gloves are recommended when collecting the mulberries to prevent staining the hands. Soak for 1-2 hours then slowly raise the temperature to nearly boiling and simmer for 30 - 60 min. Strain to remove the berries and 30 mL (2 Tbs.) vinegar to the dye bath and sufficient water to make a 4 L volume. Add wet alum mordanted wool or wet alum/tannin mordanted cotton or wet tannin mordanted cotton *without alum* and simmer 1 hr.

Remove from dyebath and rinse in afterbath made from 60 mL (¼ cup) vinegar and 1 L water. Athough any type of vinegar will work, distilled vinegar is recommended to avoid introducing other colors.

Cranberry Dye (modified from Hess)

Cranberries contain the anthocyanins cyanidin and pelargonidin (Harborne 1976). Place 200 g fresh chopped cranberries and 4 L of water into an 8 L beaker or a stainless steel or enamel pan and soak overnight. The next day bring to nearly boiling. Simmer for 30 - 60 min. Strain to remove the berries and add 30 mL (2 Tbs.) vinegar to the dye bath. Add wet alum mordanted wool or wet alum/tannin mordanted cotton and simmer 1 hr. Remove from dyebath and rinse in afterbath made from 60 mL (¼ cup) vinegar and 1 L water. Although any type vinegar will work, distilled vinegar is recommended to avoid introducing other colors. This will yield a reddish pink dye that will eventually fade to beige or tan.

Rosy Beige Herbal Tea Dyes

Purchased Celestial Seasonings Red Zinger™ or similar red colored herbal teas are the dyestuff in this recipe. The active ingredient in Red Zinger is hibiscus flower. Other red herbal teas have hibiscus, rose hips or other red flowers. In all cases the dye chemical are anthocyanins, probably including cyanidin, found in most red roses (Harborne 1976) and malvidin found in hollyhocks and related flowers (Harborne 1976). Add 15-20 tea bags to 4 L of water and simmer for 30-60 minutes until the dye bath appears to have reached a constant color. Remove the tea bags and stir in 1-2 Tbs. of vinegar. Add wet alum mordanted wool or alum/tannin mordanted cotton and simmer for 1 hr. This will produce a pale pink dye that will eventually fade to a rosy beige. More tea bags may produce a deeper shade but will add to the expense.

<u>Chamomile</u>

Chamomile is an often cited dye stuff (Androsko, Buchanan, Cannon & Cannon, Casselman, Epp, Schetsky) and many students will know about it. It produces various golden yellow shades on both cotton and wool. The plant species used by various authors differ, but the most common and reportedly best dye stuff is *Anthemis tinctoria* or dyer's chamomile (Androsko, Buchanan, Cannon & Cannon). It can be found as a wild flower called Golden Marguerite (Schetky). Common chamomile, or *Chamaomelum noblile*, is a related genus most commonly used for chamomile tea (Cannon & Cannon). Buchanan says that this species is also known by the older name *A. nobilis*. Both agree that it is suitable for a dye although it contains less of the dye than *A. tinctoria*. Casselman lists five genus and species designations for chamomile useful as a dye stuff but none of them are *A. tinctoria* or *C. nobile*.

Common chamomile or Roman chamomile, both *C. nobile*, can be grown from seed or purchased herb plants but I have not had luck with them in the poorly draining clay soil of southwest Ohio. The plants grow and produce a modest number of flowers, but often die back during the hot dry part of the summer. It may grow better in other geopgraphical locations.

The active compounds in *A. tinctoria* include the flavones luteolin and apigenin (Fig. 6.11a and b), and the flavonols quercetagetin and patuletin (Fig.6.12b and f). Luteolin and apigenin are also found in the yellow dye obtained from Weld, *Reseda luteola* (Cannon & Cannon, Epp, Mayer & Cook, Sequin-Frey). The flavonols quercetagetin and patuletin are also an important dye chemicals in Marigold as discussed in Chapter 3.

Dyer's chamomile is available from specialty dye sources, or can be grown as a plant in a dyers garden. Another source of chamomile dye stuff is chamomile tea bags from the grocery store, but check the ingredients. Some chamomile teas also contain other herbs which may affect the colors. Celestal Seasonings™ sells a chamomile tea which is pure chamomile. Most sources either do not specify the fabric or use wool only. We have used chamomile from tea bags to dye both cotton and wool with and without alum/tannin or alum as a mordant. The color obtained with alum/tannin on cotton is more intense than unmordanted cotton. Samples of fiber dyed with chamomile from tea bags are shown in photo 5a in the color insert section of the book.

Chamomile Yellow Dye Recipe

Use chamomile flower heads or purchase chamomile tea bags. Not all brands of chamomile tea are pure chamomile, so read the ingredients list! Add 1 L of flower heads or 15-20 tea bags to 4 L of water and simmer for 30-60 minutes until the dye bath appears to have reached a constant color. Strain to remove the flowers or remove the tea bags from the bath and cool slightly. Add wet alum mordanted wool or wet alum/tannin mordanted cotton, return to nearly boiling and simmer for 1 hr.

Coffee

Coffee, *Coffea arabica,* can also be used as a dye. Because coffee is expensive leftover coffee grounds from coffee makers or used "coffee bags" are the most frequently cited sources for the dyestuff. These can be collected, dried out and saved for later use. Leftover coffee in the pot from the office is sometimes suggested as a source of dyestuff but is more difficult to store. Instant coffee is undoubtedly the most convenient source and the source recommended here. This recipe for coffee dye is intended for small amounts of fiber and small pieces to be "antiqued." The cost is not prohibitive in such small amounts. Coffee contains tannins and probably flavonoids that serve as coloring agents. The brown color in tea and chocolate is caused by oxidation of hydroxyflavans (Swain 1976) and this may account for the brown of coffee as well.

Instant Coffee Dye for "Antiquing" Small Quilts and Cotton Doilies

Dissolve 3-4 Tbs. of instant coffee in one cup boiling water. Pour the coffee solution into a 4 L beaker or large bowl and dilute with hot (not boiling) water to 2 L. Immerse the item to be antiqued into the solution and let stand 15 - 60 min until desired color is obtained. The color will appear darker on wet fabric so wait until the item looks slightly darker than you want it to be. This method will yield a brown antiquing that is less orange than the tea antiquing recipe given below. Remove and rinse. If it appears too light it can be returned to the dyebath. These items will not be laundered frequently and so it is not necessary or advisable to use a mordant. They can be re-dyed if the color fades with repeated laundering or light exposure.

Instant Coffee Dye for Fibers.

Follow the recipe above but add ½ to 1/3 cup of instant coffee powder to a 4 L beaker or stainless steel or enamel pan and bring to nearly boiling. Add wet alum mordanted wool or wet alum/tannin mordanted cotton and simmer for ½ - 1 hr.

Mint

Mint species like peppermint, *Menthus piperita*, and spearmint, *Menthus spicata*, are often grown by herb gardeners. Anyone who grows mint has more than s/he can use and will probably be willing to give some away. It is easy to grow and is very hardy. It spreads rapidly and can become an intruder if not weeded frequently. Most herbs prefer well drained soil and full sunlight, but mint grows well in places where many plants will not. I have my mint at the edge of my driveway, a rocky location with little topsoil and sun for only part of the day. It grows well and is prevented from invading other garden beds.

Mint can also be purchased as mint herbal tea, although more expensively. Purchased mint herbal tea is usually a mixture of peppermint and spearmint and some other plants. Pure peppermint tea is available from Bigelow™ and many suppliers sell mixed mint teas. Mint dyes produce a nice taupe color probably as a result of tannins (fig 6.15) and flavonoids. The browning of fruit is caused by oxidation of hydroxyflavans (Swain 1976) and this may account for the brown tones in mint dye as well. The fresh mint produces a color that is richer and deeper than can be obtained from home dried mint or mint tea bags. A samples of fabric dyed with mint is shown in photograph 6a in the color insert section of the book.

Fresh Mint Dye

Place about 1 L of chopped or torn mint leaves and stems in a 6-8 L beaker or large stainless steel or enamel pan. Add sufficient distilled water to easily cover them to a depth of several inches. Bring to a boil and simmer 30-60 min. stirring occasionally. Strain with wire mesh to remove the mint plants. Add water if needed to make a dyebath of 3-4 L. Add wet fiber and heat to nearly boiling, then simmer 30-60 min. This dye will work on unmordanted fibers, but may produce a deeper color on alum mordanted wool or alum/ tannin mordanted cotton. The resulting color is beige with dried mint and khaki brown with fresh mint. Mint can be dried for later use, but as is the case with the tea bags, the dye color is not as rich as that with fresh mint.

Mint Herbal Tea Bag Dye

Use purchased mint tea bags. Read the ingredients list to be sure the tea is pure mint tea. Other ingredients may change the color of your dye. Add 15-20 tea bags to 4 L of water and simmer for 30-60 minutes until the dye bath appears to have reached a constant color. Remove the tea bags from the bath and cool slightly. Add wet fiber and return to nearly boiling and simmer for 1 hr to yield a tan dye. This dye will work on unmordanted fibers, but may produce a deeper color on alum mordanted wool or alum/tannin mordanted cotton.

Tea

Tea, *Thea sinensis*, can be used as a dye. The trick of dyeing stained household items with tea dye was well known to our foremothers. Home decorating magazines will sometimes suggest tea dyeing new doilies, table scarves and linens to make them look like vintage finds. Quilters interested in antique quilts often "tea dye" new quilts to soften the colors and give them the brown patina of antiques they are reproducing. Tea bags can be saved after brewing tea for drinking and later used to dye, or inexpensive tea bags can be purchased specifically for dyeing. Black tea is reported to make the best dye (Casselman, Lesch) and fortunately is found in larger amounts in less expensive "generic" brands of tea. If you are purchasing tea bags specifically for dyeing, the bargain brands usually make fine dyes.

Tea contains as much as 30% tannin (Hendrickson) and this is undoubtedly one of the dye chemicals and a mordant. Instant tea powder also contains quercetin (fig 6.12a) and related flavonols (Cannon & Cannon) and presumably these are also present in the tea leaves found in tea bags. They are most likely responsible for the orangish color produced by tea dyeing. The brown color of tea is caused by the oxidation of hydroxyflavans (Fig. 6.21) (Swain 1976). Because tea dye is frequently used to "antique" small quilts or cotton doilies, a recipe is given below. Samples of fibers dyed and antiqued with tea are shown in photograph 6b and c in the color insert section of the book.

Tea Bag Antiquing Recipe

Place 8-10 tea bags in a glass, stainless steel or enamel bowl or pan. Add ½ L of boiling water and steep tea for 5-10 minutes or until deeply colored. Remove tea bags* and dilute dyebath to a volume of 2 L with hot (not boiling) water. Immerse item to be antiqued into the dyebath and let stand, stirring occasionally for 15 min – 1 hr. depending on final color desired. This will yield an orange-brown color, more orange than coffee antiquing recipe above. The wet cotton will look considerably darker than when dried. Some authors make an antiquing dye by mixing coffee and tea dyes. This generally yields a darker brown color with less orange cast.

*Tea bags may be left in the dyebath during the dyeing process and will produce a mottled look on the fabric which some people prefer.

Tea Bag Dye Recipe

Purchase inexpensive tea bags and add15-20 tea bags to 4 L of water and simmer for 30-60 minutes until the dyebath appears to have reached a constant color. Remove the tea bags from the bath and cool slightly. Add wet fiber and return to nearly boiling and simmer for 1 hr. This dye will work on unmordanted fibers. Tea dyes produce a range of orangy-brown colors. Teas that contain more black tea will reportedly produce a darker dye while those with more orange tea produce an orange brown dye.

Weld

Weld, *Reseda luteola*, is an ancient European dye that has been found in Neolithic villages in Europe (Cannon & Cannon). It was widely used as a yellow dye throughout European history including the wedding gowns of ancient Rome and the yellow caps Jews were forced to wear in the middle ages (Cannon & Cannon). It was also used by some African tribes (Brunello).

Weld can be grown from seed (Buchanan, Cannon & Cannon). The seeds can be hard to find and the plant may not sprout the first year, but once established it will readily re-seed itself. One planting will produce large amounts of weld that can be cut off at the base and hung upside down to dry. I obtained weld seeds through a seed exchange program of the Herb Society of America and produced enough plants for several years of use. Smaller plants did sprout in the same location the next year from self sowing. The dye chemicals in weld are similar to those in chamomile, namely luteolin (fig 6.11a) and apigenin (fig 6.11b) (Cannon & Cannon). Samples of fiber dyed with weld are shown in photograph 5b in the color insert section of the book.

Weld Yellow Dye Recipe (method from Buccigross, Jill Goodwin, Wiegle)

Chop or break up about 120 g fresh or dried weld leaves and stalks into an 8 L beaker, a stainless steel or an enamel pan. Add sufficient distilled water to easily cover them to a depth of several inches. Dried weld may benefit by being soaked for several hours or overnight to rehydrate it before beginning the simmering process but this is not necessary with fresh plants. Bring the pot to nearly boiling and simmer 45-90 min. stirring occasionally. The dye bath is often described as having the odor of cooking asparagus and this is true if fresh weld is used. Strain with wire mesh to remove the weld. Add wet alum mordanted wool or wet alum/tannin mordanted cotton fiber to the dye bath and heat to nearly boiling. Simmer 60-90 min. The resulting color is yellow to greenish yellow. In my experience, the color appears richer and deeper when fresh weld is used. Jill Goodwin reports that doubling the amount of weld will produce a greener dye but I have not tested this.

Flower Pounding

A related technique to dyeing is the traditional folk art of "flower pounding." In this process, mordant treated fabrics are selectively dyed by arranging flowers or flower petals onto the fabric and then them smashing them to release the dye. Interestingly, some of the plants which make the best dyestuffs are less satisfactory when they are pounded because they produce too much dye and the pounded image is a smudge of intense color rather than an image of the petal.

This is a fun and interesting technique that can be used with leaves and flowers that have less dye chemicals. It would be an interesting activity for a biology class studying botany. It is also a convenient way to quickly audition new flowers and leaves as potential dye candidates. Pounding blue flowers will very quickly illustrate the "what you see isn't necessarily what you get" aspect of dyeing. The students will be surprised to see that pale yellow images usually result from blue flowers. Similarly, rinsing images produced by anthocyanin or anthochlors pigmented flowers in afterbaths of vinegar or baking soda will cause color changes that can lead to a discussion of their chemistry (see Ch. 8).

Some students will be attracted to flower pounding because it allows more artistic arrangement of the colors than dyeing alone. Some crafters choose to make "wallpaper" like designs with petals and leaves while others arrange their petals carefully to produce images that look like flower bouquets. Some leave them natural and others outline and highlight the pounded images with permanent fabric markers. A basic technique is given here, but there are many variations and there are several flower pounding books available in bookstores. The interested reader can easily find these by entering "flower pounding" in the subject area of any on-line book dealer.

Flower Pounding Method

The supplies required are alum/tannin treated cotton or similar fabric, a large wood cutting board (that will not be used for food) or a scrap of plywood, and a hammer. Plastic wrap or wax paper and transparent tape can be helpful if you want to arrange the flowers carefully. If desired, the images may be outlined and highlighted to make them look more like flowers. In that case, fine point permanent fabric markers such as Pigma Pens™ or Gel pens that are suitable for fabric are also required.

Choose the flowers and leaves to pound and separate them into individual petals and leaves. Place the mordanted fabric, good side up, on the scrap wood or old cutting board. Arrange the flower petals and leaves on the fabric as you want the final image to appear. Several petals can be placed at once and taped to the fabric to keep them from shifting or petals can placed one at a time and pounded before placing any others. Once the petals and leaves are arranged, cover the fabric and leaves with plastic wrap or waxed paper. Now pound the flowers pieces with the hammer. This makes a lot of noise and takes harder pounding than might be imagined so it's a good idea to do it outside. Remove the plastic wrap and scrape off as much of the flower material as possible. Continue working on the design until completed.

Some crafters choose to use a second piece of mordanted fabric on top of the image in place of the plastic wrap so two mirror images can be pounded simultaneously. This variation will not work well if the flower petals are taped in place because the dye will not penetrate the tape. It may also result in paler images because the same amount of dye is now being used to dye two pieces of fabric. The major disadvantage to this adaptation is that the petals and leaves are no longer visible when pounding. It is difficult to know if they shift and can be hard to know when dye transfer is complete. Wax paper or plastic wrap is recommended until the dyer has had some experience with this process.

After pounding, let the flower dyed fabric dry completely. Once dry the flower shapes can be outlined and highlighted with permanent fabric markers and small details added to make pounding as artistic as desired. Many crafters use these pounded images to make pillows. It would be an interesting activity with younger children to make a pillow by pounding one panel on fabric and dyeing additional fabric with the same plants to serve as borders and/or backing for the pillow. These can be great gift ideas for students to give to parents and can be easily worked into an integrated curriculum in elementary and middle school classrooms.

Part 3
The Science of Dyeing

There are many important factors in making something a good dye. The chemical must absorb visible light, it must be soluble in water or be capable of being made soluble and it must form a bond with the fabric or with a mordant in the fabric. A good dye should be stable to washing and light exposure. An understanding of the chemistry of dyes and dyeing must begin with an understanding of the physics of light and color and the chemistry of the molecules that absorb and reflect that light.

Chapter 6
The Physics and Chemistry of Light, Color, and Dyes

Light and Color

Issac Newton was probably the first person to recognize that the visible spectrum produced by a prism, and the familiar colors of the rainbow, "ROY G BV", or red, orange, yellow, green, blue, and violet, are all components of white light (fig. 6.1a). Now scientists know that visible light is part of the much larger electromagnetic spectrum. Visible light is simply the narrow band of wavelengths that can be perceived by the human eye (fig. 6.1 b). This visible region is usually listed as 400-700 nm but wavelengths as short as 360 nm and as long as 780 nm can be observed at high light intensities (Christie, Nassau a).

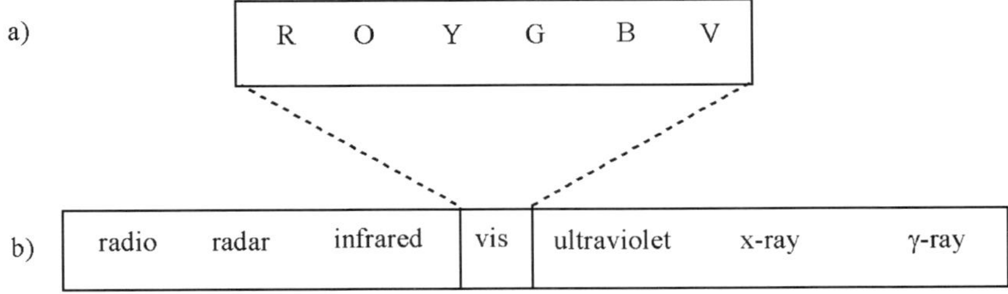

Figure 6.1 a) The visible spectrum of light. Different colors of light have corresponding wavelengths and frequencies as described in the text. b) Visible light is only a small part of the electromagnetic spectrum. (Nassau a).

Most of the colors observed in daily life, however are not produced by passing white light from a luminous source through a prism. When viewing opaque physical objects they are visible because of the light they reflect rather than light they produce. If an object is viewed in white light and reflects all the colors of light equally well, it will appear white. Similarly, an object that reflects all the colors of light equally poorly will appear black because the black object has absorbed the light. Color is what happens when an object absorbs some colors of light more than others. There are several ways in which a colored object can result. In the simplest case, an orange object may be orange because light in the narrow band around 600 nm is reflected and all other light absorbed (Nassau a). In real life, however, the situation is rarely this simple. Orange paints absorb only the short wavelength end of the spectrum, the V, B and G light and reflect the wide band of Y, O and R wavelengths yet appear orange (Marcus).

Of more interest to us is the mechanism of color production in dyes. To be bright, dyes must absorb only a narrow range of wavelengths (Christie). A bright orange dye would absorb light in the greenish-blue region. In this case the light reflected to the eye is deficient in only one color of light, greenish-blue, and reflects the rest of the visible spectrum to the eye. The color observed however does not appear to be a blend of colors, but the compliment of the absorbed light, as if only one color of light were reflected (Christie, Griffiths). This effect is shown in figure 6.2. The situation can be even more complicated if mixes of dyes or paints with different absorbance ranges are used to produce a color (Nassau a).

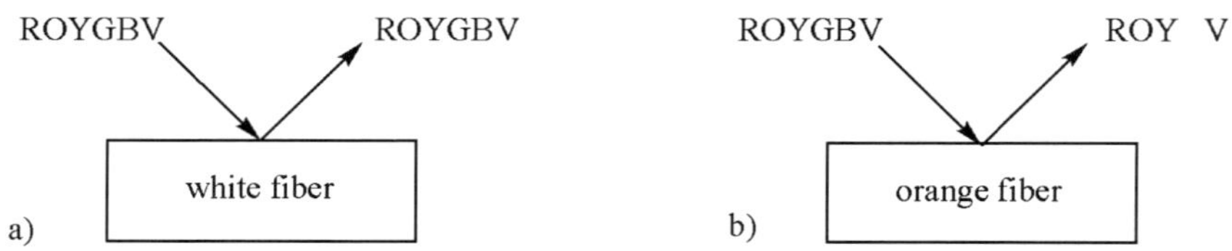

Figure 6.2 A white fiber reflects all colors of visible light equally. A colored fiber will reflect most colors and absorb one. The perceived color of the fiber is the color of only part of the light that is reflected. The perceived color is determined by the color absorbed rather than the sum of those reflected.

While it is not intuitively obvious how absorbing one color of light and reflecting several results in the eye perceiving only one color, it is easy to predict with a color wheel. The traditional artists color wheel is a circle segmented into six parts, with each segment assigned to one spectral color. Traditionally physicists construct the wheel by beginning with red at the top and proceeding clockwise in order of decreasing wavelength (fig. 6.3a). Simplistically, the color observed is the hue that is directly opposite the absorbed color on the color wheel. In color theory terms, the reflected color is called the compliment of the absorbed color because together a colored pigment and its compliment will absorb all the colors of light nearly equally to produce a black, gray or muddy brown (Griffiths, Trotman). The simple complimentary color method using the color wheel is a simplification because in reality the color "wedges" are not of equal size. An improved complimentary scheme can be constructed using ten segments as shown in fig. 6.3 b. This expanded color wheel more correctly predicts observed colors.

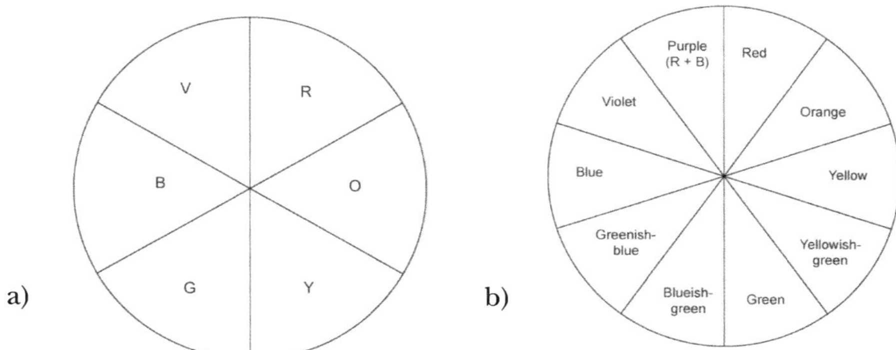

Figure 6.3 The color wheel is used by artists to understand colors and color theory. The compliment of any color is the color opposite it on the color wheel. A color of light reflected by an object should be the compliment of the color it absorbs. a) The simple 6 segmented color wheel is shown here (Trotman). b) An improved 10 segment color wheel more correctly predicts observed color from absorbed color (Griffiths). Purple is a non-spectral color composed of blue and red light.

Another complication arises when using the complimentary color method with real objects (Simon). Real dyes and pigments are not ideal or optimal colors. An ideal green object would reflect to an identical extent all the wavelengths of light that constitute green. It would completely absorb all other wavelengths of visible light. Although some optical filters approximate the ideal, nature does not. Ideal colors are theoretical constructs (Simon). By contrast a real green object has its maximum absorption in the wavelengths of light that constitute the compliment of green but does not absorb all these wavelengths completely or necessarily to an equal extent. Neither does it totally reflect all other wavelengths of visible light. A theoretical reflectance curve for an ideal or optimal green object is shown in figure 6.4a below and a reflectance curve for a real green object is shown in fig 6.4b.

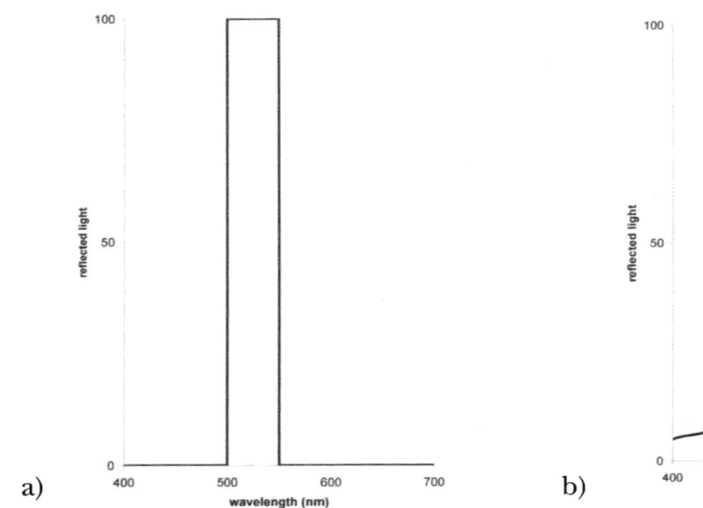

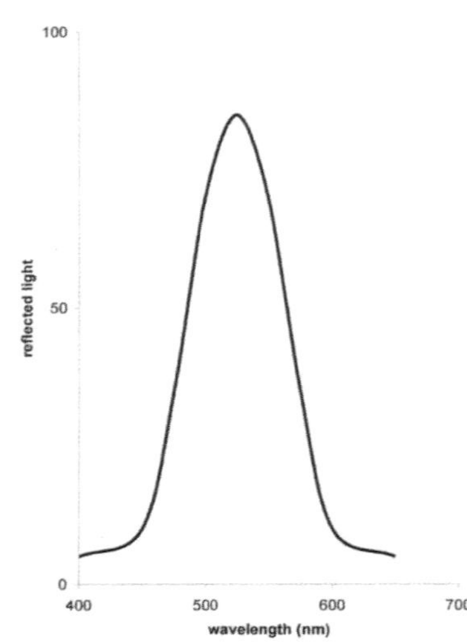

Figure 6.4 Reflectance curves a) The theoretical reflectance curve for an ideal or optimal green object. Ideal colors are theoretical constructs and do not exist in nature. b) A reflection curve for a real green object. Real colors simulate ideal colors.

The absorbed light and the resultant reflected color can be described more accurately by considering the actual wavelengths of light absorbed rather than just its spectral color. There is some disagreement among various sources as to what wavelengths constitute a particular color (see for example Nassau a), but the differences are relatively minor. This quantitative treatment more accurately predicts the color observed when a narrow band of light is absorbed as shown in Table 6.1 (Christie, Griffiths, Trotman). As can be seen in the table, not all colors contain an equal range of wavelengths. This differs conceptually from dividing the spectrum into equally sized wedges as is done with color wheels.

Table 6.1 The color of an object depends on the wavelengths of light it absorbs. The wavelengths and colors of light absorbed and the observed color of the object are shown. This table is adapted from Trotman.

Wavelength absorbed (nm)	Color of light absorbed	Color of light reflected
400-435	violet	yellowish-green
435-480	blue	yellow
480-490	greenish-blue	orange
490-500	bluish-green	red
500-560	green	purple (red + blue)*
560-580	yellowish-green	violet
580-595	yellow	blue
595-605	orange	greenish-blue
605-750	Red	bluish-green

*Purple is not used as a synonym for violet. Here violet is regarded as a spectral color of specific wavelengths whereas purple is used to mean a color produced when both red and blue light are observed simultaneously.

The table highlights several interesting points. First it is clear that absorbing a narrow range of wavelengths will result in one color being observed even though many other colors of light are also present. As described above, the observed color is the compliment of the light absorbed and not a blend of colors. The reasons for this have to do with the physiology of the eye rather than the nature of the spectrum. There are three different sensitivities in the receptors of the human eye for long wavelength or red, medium wavelengths or green and short wavelengths or blue regions of the spectrum (Christie, Griffiths, Krauskopf). These are the "additive primaries" or "light primaries." When a "color" of light is removed, the receptor for that color in the eye is not triggered but receptors for the other two primaries are. The brain interprets the color of the object by using the rules of additive color mixing (Griffiths). The resulting color is the compliment of the missing light. Readers interested in this fascinating area of physiology or the details of additive color mixing are directed to the references above.

Second, the table shows that sometimes the observed colors can differ significantly from the qualitative predictions of the simple six segment color wheel of fig 6.3a. The artists color wheel predicts that red will be observed when green light is absorbed, but in reality red is only one of three colors that can be observed when green light is absorbed. Which of the three colors is seen depends upon what particular wavelengths of green light are absorbed. Table 6.1 shows that absorbing green light from the blue end of the green spectral range (bluish-green) does result in a red color being observed. However, absorbing light in the mid green range results instead in a purple color being seen. Finally, absorbing light from the yellow end of the green range (yellowish-green) produces a violet color. A similar effect can be seen with blue light absorption. The ten segment color wheel (fig 6.3b) includes more divisions for these blue and green spectral regions and correctly predicts the observed color and the physics compliment. It is essentially a qualitative version of the table.

A third interesting point that can be seen in table 6.1 is that no single spectral color is a compliment for middle green. To observe the color green both red and blue wavelengths of light must be absorbed (Griffiths, Trotman). Green plants contain the green pigment chlorophyll. Chlorophyl absorbs red light of 660 nm and blue light of 430 nm producing a compound that appears green (Griffiths). The absence of good natural green dyes has been reported by many authors. This requirement to have light absorbed in two different wavelength ranges is responsible for the shortage of green dyes. To understand the difficulty of a dye absorbing light in two different wavelength ranges requires some understanding of how and why dyes and other substances absorbs light.

Why colored objects absorb certain colors of light has to do with the nature of matter. All matter is composed of atoms and molecules. The electrons in atoms and molecules can absorb light when the energy of the light is exactly equal to the energy spacing between the electron energy levels, or orbitals in the atom or molecule. In other words, when the energy gap or step is exactly equal to the energy of the photon of light, the light will be absorbed and the electron will jump to the higher energy level. This is called an electronic transition (Fig. 6.5).

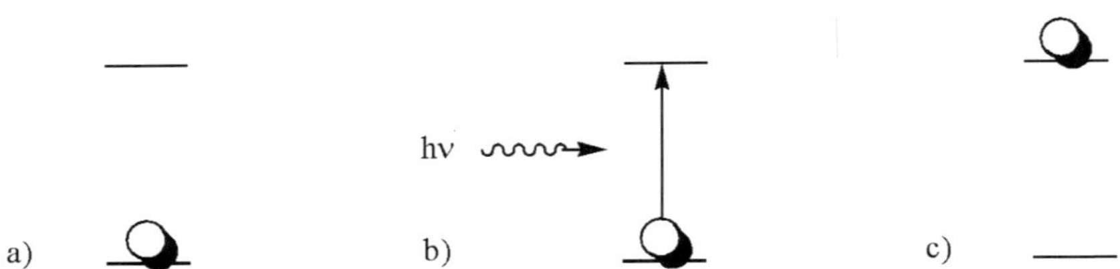

Fig. 6.5 An electron will absorb light if the energy difference between two electron energy levels or orbitals is exactly equal to the energy of the light. a) an electron in the ground state of the atom or molecule. b) a photon of light with energy exactly equal to delta E is absorbed. c) the electron in an excited state. This excitation of an electron from a ground state to an excited state by absorption of light is called a transition.

The energy of light, or any radiation, depends on its frequency according to eq 1 where E is energy in Joules, nu is frequency in inverse seconds, or Hz and h is Planck's constant. High frequency radiation then, is high energy radiation.

$$\Delta E = h\nu \qquad \text{eq 6.1}$$

Most people are more accustomed to thinking of colors of light in terms of wavelengths than frequencies. The frequency and wavelength of light are inversely related according to eq 2, where c is the speed of light in m/s, lambda is the wavelength in meters and nu is the frequency in inverse seconds, or Hz. A long wavelength of light will have a low frequency and thus according to eq 1, low energy as well.

$$c = \lambda\nu \qquad \text{Eq. 6.2}$$

In the visible spectrum, red light is the lowest energy light and will be absorbed when the energy levels are relatively closely spaced resulting in a bluish-green object. At the other end of the visible energy spectrum, violet light will be absorbed by molecules whose energy levels are more widely separated resulting in a yellowish green object. As absorption proceeds through the visible spectrum from red to violet, the observed colors will not follow the inverse of the spectrum but rather the order of the compliments of the spectrum. They proceed from bluish-green through to yellowish-green as seen in Table 6.1. As mentioned before, there is no spectral color whose absorbance will produce green. In molecular terms, a green dye molecule would need two different electronic transitions in the visible range, one red and one blue. The range of energies represented by visible light is very small. Few molecules have energy levels spacing required to produce the two different absorptions in the visible region.

The Chemistry of Dyes

The absorption of light by electrons in molecules relates directly to the chemistry of dyes and dyeing. A piece of white cotton fabric or yarn appears white because the cellulose molecules of the cotton reflect all the colors of light. When the fiber is dyed the molecules of the dye chemical are attached to the fiber and are viewed along with the fiber. These dye molecules absorb some of the wavelengths of white light that fall on the cotton. The reflected light that reaches the viewer's eye now has been depleted of a range of wavelengths and the cotton takes on the complimentary color of the absorbed light. But what makes a Dye Chemical a Dye?

There are many compounds that are not colored. All compounds can absorb electromagnetic radiation. This is particularly true of organic compounds, which usually absorb radiation of many different wavelengths. To be a dye, however, the chemical must absorb radiation in or very near the visible region of light, a very small part of the electromagnetic spectrum as seen in Fig. 6.1 b. The question of what makes something a dye then becomes a question of what type of molecule will absorb in the visible region of the spectrum.

Historically the search to answer that question began by looking at the structures of known dye chemicals. As was true in most of early organic chemistry, understanding dyes was a task of categorization. Certain functional groups appeared in a large number of dyes. Witt classified these groups as "chromophores" in the 1870's and 1880's (Christie, Swain 1976). They were believed to be responsible for producing color in dyes because chemical reduction of the groups destroyed the color of the dyes containing them (Christie, Trotman). Although the word chromophore has persisted in dye chemistry, no chemical definition of a chromophore has been universally agreed upon (Griffiths). In 1928, Dilthey and Witzinger noted that Witts chromophores were all electron withdrawing (electron poor) groups and proposed that definition. More recently Griffiths defined a chromophore as an unsaturated functional group, that does not produce color *itself*, and that definition will be used here. The chromophores found in natural dyes include nitro, nitroso, azo, carbonyl groups and quinoid rings (Christie, Padgett) shown in Fig. 6.6.

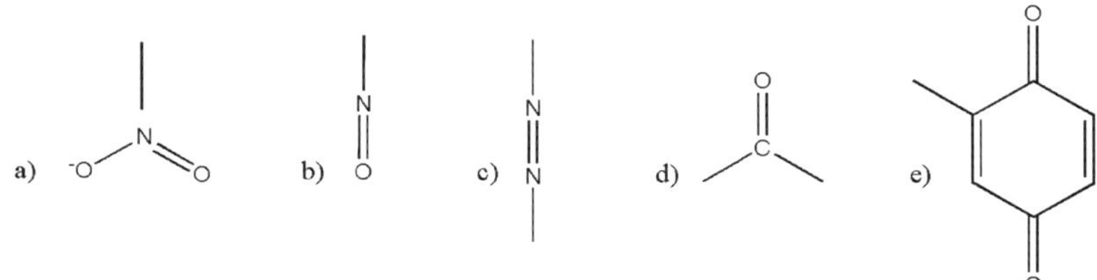

Figure 6.6 Chromophores or functional groups found in a large number of dyes were catalogued. They include a) nitro b) nitroso c) azo d) carbonyl and e) quinoid.

The topic was more complicated because additional common functional groups, or "auxochromes" were present in dyes. Witt coined the term auxochrome and defined it as groups that deepened the color of dyes (Christie, Swain 1976). Sometimes changing the auxochrome would change the dye color (Trotman). While auxochromes were categorized, they were not chemically defined either until 1928 when Dilthey and Witzinger discovered that Witt's auxochromes were all electron donating groups. This definition is still appropriate today and recently Griffiths defined an auxochrome as an electron donating group. This definition is particularly useful based on the current understanding of dye chemistry and so will be used here. Auxochromes in natural dyes include hydroxyl, O-alkyl and amine functional groups (Christie, Griffiths, Padgett).

The picture was still not complete because there are molecules with no chromophores or auxochromes that are colored. Graebe and Liebermann recognized that unstauration (double bonds) were important for color in 1868 even before Witt's chromophore/auxochrome system was proposed (Christie, Swain 1976). Finally in 1907 Hewitt and Mitchell recognized that dye molecules always have long segments of alternating carbon-carbon single and double bonds (Christie, Griffiths, Swain 1976, Trotman). This pattern of alternating double and single bonds is called "conjugation." A conjugated chain with 8 or more double bonds is colored (Griffiths). Lengthening the conjugated chain of a dye changes the color of the dye (Christie, Swain 1976, Trotman). To understand why conjugation is important to dyes we need to understand conjugation. Historically the understanding of conjugation began with benzene.

Benzene was believed to be a cyclic structure with alternating carbon-carbon single and double bonds as shown in Fig 6.7 a. The problem with benzene however, was that it didn't behave as if it had both single and double bonds. Everything seemed to indicate that all six carbon-carbon bonds were identical. But that wasn't the only problem. Benzene was much more stable, with much lower energy than would be expected compared to other carbon compounds. Benzene can be drawn with two different arrangements of double bonds as shown in Fig 6.7 b. The explanation of how single and double bonds can behave as though they are identical came to Friederich Kekule in 1865. Benzene, he believed existed in both forms and fluctuated rapidly between the two structures. This fluctuation was called resonance and the two different structures are known as resonance structures. So, according to Kekule, each bond would sometimes be single and sometimes be double, and thus all six bonds were identical.

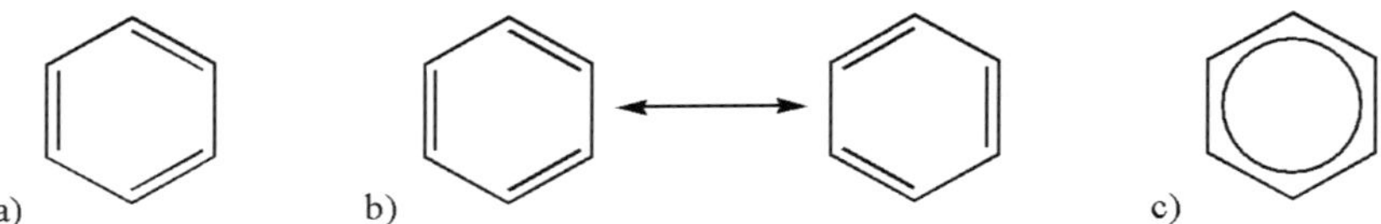

Fig 6.7 Benzene has resonance structures. a) The single benzene structure does not accurately represent benzene's chemical properties. b) Resonance structures approximate benzene's structure more correctly. c) Benzene is often drawn with a circle representing the delocalized electrons as explained in the text.

There are many organic compounds with conjugation and resonance. Some of these had many more than two resonance structures possible. Rules were drawn up to explain what made good and bad resonance structures and an idea called resonance theory was born. The reason for benzenes unusual stability was not explained until 1931 when Hückel showed that having odd numbers of conjugated double bonds, confers molecules with extra stability. Molecules with this arrangement of electrons are referred to as "aromatic." Aromatic compounds are more stable and lower in energy than similar non-aromatic compounds. Many natural dyes are aromatic as well as being conjugated (Christie). Both resonance theory and aromaticity were applied to dyes and dye chemistry.

In 1935 Bury noted a relationship between dye color and resonance (Christie). In resonance theory the fluctuation of conjugated systems between the various resonance structures was thought to be responsible for the observed properties of conjugated systems, including the color of dyes. In many cases, resonance theory gives good qualitative results for dyes (Christie) but has not proven entirely successful in explaining dye color (Griffiths). The idea of rapid fluctuations of molecules between resonance structures is not longer accepted and cannot be justified by a modern understanding of chemistry.

Eventually it was proposed that the true form of conjugated molecules, including dyes, was a merger or hybrid of the resonance structures. In this theory, electrons in conjugated double bonds are "delocalized," that is, they are not confined to the space between two atoms but belong to the whole molecule. Thus, in a sense, the pattern of alternating double and single bonds is not real and all the delocalized electrons belong to all the atoms in the conjugated system. To more accurately reflect this idea, delocalized electrons are sometimes represented as a circle, rather than individual localized bonds. This is shown for benzene in figure 6.7 c. It appeared that having delocalized electrons made a molecule a good dye and that explained the relationship between resonance forms and dye colors.

A new understanding of delocalization of electrons was made possible with the advent of quantum mechanics. With quantum mechanics came the idea that *all* electrons, not just the delocalized electrons, belong to the entire molecule. The fact that some electrons seem to be fixed between two atoms is simply a matter of probability and the type of orbital they occupy. Some orbitals are confined to part of the molecule; others are spread over the entire molecule. The idea that molecules have "molecular orbitals," rather than a series of atomic orbitals on individual atoms, is known as Molecular Orbital Theory.

In Molecular Orbital Theory a bond is formed between two atoms when the atomic orbitals (ao) of individual atoms overlap and are combined to form new molecular orbitals (MO). While the MO always belong to the entire molecule, not all electrons in all MO are delocalized. It depends upon the type of molecular orbital that is formed. End to end overlap of atomic orbitals results in the formation of sigma molecular orbitals, also called sigma bonds as shown in fig 6.8 a. The electrons in sigma bonds are usually localized between two atomic nuclei. Another type of MO is formed when atomic orbitals interact by side to side overlap. This type of overlap produced pi orbitals or pi bonds as shown in fig. 6.8 b. The electrons in pi bonds are usually delocalized electrons.

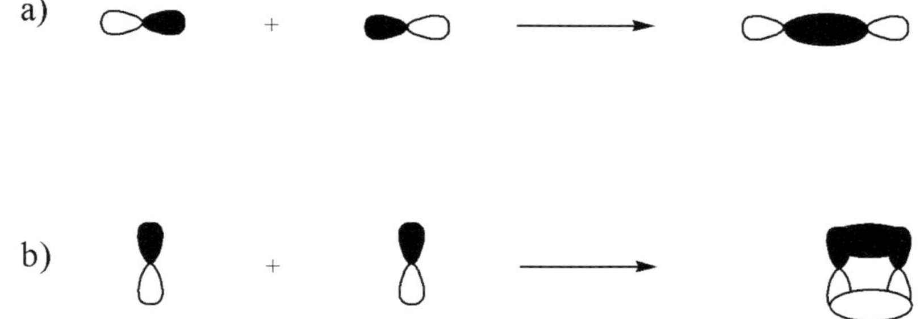

Fig. 6.8 Molecular orbitals are formed by the overlap of atomic orbitals. a) End to end overlap of atomic orbitals produces localized MO called sigma bonds. b) Side to side overlap of atomic orbitals produces delocalized MO called pi bonds.

The overlap of two atomic orbitals will result in two MO, one of lower energy than the original ao and one of higher energy than the ao. This will be true no matter how many aos are combined to produce MO. The rule is that there will be the same number of MO as ao. Usually half the MO will be lower in energy than the ao they were formed from and half higher in energy. The MO with lower energy are called bonding orbitals. Electrons in bonding orbitals stabilize a molecule. The MO with higher energy are called antibonding orbitals. Electrons in antibonding orbitals destabilize a molecule. Antibonding orbitals are designated with an asterisk. In the case of an odd number of orbitals, at least one orbital will be a nonbonding molecular orbital designated by the symbol n. Electrons in nonbonding orbitals usually have the same energy as the original ao and neither stabilize nor destabilize the molecule. Diagrams representing the energy profile resulting from the formation of bonding, nonbonding and antibonding orbitals are shown in figure 6.9. The electronic transitions that produce color are often from n to pi* (pi star) or pi to pi* (pi star) orbitals (Griffiths).

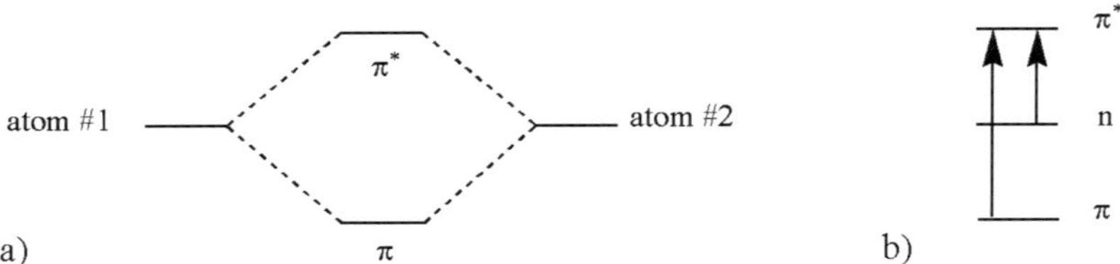

Fig. 6.9 Molecular orbitals are formed from the overlap of atomic orbitals. a) Both bonding and antibonding orbitals will be formed from the overlap of two atomic orbitals. b) Some molecules also have nonbonding orbitals (n). Electronic transitions that are common in dyes are shown.

As chemistry and chemical knowledge evolved, the question of what makes a chemical a dye entered the realm of empirical quantum mechanics and molecular orbitals. The emphasis now was not on categorizing the properties of functional groups, but looking at the molecular orbitals to understand why a dye was colored. Benzene and many other organic compounds with conjugation and delocalized electrons are colorless. A long conjugated chain is important for a dye because both conjugation and chain length affect the energy of the molecular orbitals, and thus the frequency of radiation of the transition. Lengthening of the conjugated system results in more pi and pi* molecular orbitals because there are more atoms. The more pi to pi* transitions that are possible the closer they will be in energy (Trotman). If the conjugation is long enough some transitions will be in the visible region (Griffiths, Nassau b, Swain 1976).

Molecular orbital theory as described here, began simply with sigma and pi molecular orbitals produced by the combination of two atomic orbitals, but soon involved the combination of more atomic orbitals. It was not long before computational systems were developed to allow delocalization of pi electrons. Molecular orbitals could be constructed using many atomic orbitals by the LCAO method, or Linear Combination of Atomic Orbitals. Solutions were obtained by hand using matrices and many assumptions were needed. Consequently in these early methods, the sigma bonds were considered as localized and not treated as part of the system. The pi orbitals were constructed from the p atomic orbitals of all the atoms in the pi system allowing for delocalization of pi electrons. Hückel's qualitative method for explaining aromaticity was one of the earliest attempts to do this. Having an odd number of conjugated double bonds corresponds to having (4n+2) pi electrons (where n is an integer). This is known as the Hückel rule for aromaticity. Computer modeling of

LCAO methods soon followed and even today dyes are still often modeled with Pariser-Pople-Parr, or PPP Method (Christie, Griffiths). This method has some flaws, particularly in neglecting the contributions of sigma electrons to the color, however it gives good results without requiring large amounts of computer time (Christie).

Looking at dyes from the perspective of molecular orbitals also explained why the functional groups called chromophores are important in dyes. There are two different ways in which chromophores contribute to color in molecules. In simple colored molecules chromophores participate in n to pi* transitions (Griffiths). The chromophore functional groups all contain heteroatoms with lone pair electrons, such as carbonyl oxygens, azo nitrogens and nitroso oxygens. In the molecular ground state, these lone pairs are not involved in the delocalized molecular orbital system. They occupy atomic orbitals on the heteroatom called n molecular orbitals.

The electronic transition that produces color in n to pi* systems occurs when a lone pair electron is promoted from the n orbital to the lowest empty molecular orbital, usually a pi* orbital. In quantum mechanical terms, the transition consists of the transfer of electron density from the heteroatom of the lone pair into the delocalized molecular system (Grffiths, Nassau b). These transitions usually have energies that correspond to the visible region of the electromagnetic spectrum. The blue color of nitrosomethane and the pale yellow color of nitrobenzene can be explained by n to pi* transitions (Griffiths) as can the color of many other small organic molecules. Most dyes however, are not small molecules and although some dye molecules do have n to pi* transitions, they are not the strong transitions that account for the dye colors produced. Those are produced by a more complex process of electron density transfer.

Most natural dyes can be classified as donor-acceptor systems (Griffiths). In these systems, the electronic transition involves the transfer of electron density from an auxochrome, the donor, to a chromophore group on the molecule or to the molecule as a whole (Griffiths, Nassau b), the acceptor. In the molecular ground state of a donor-acceptor compound the donor group is linked to the conjugated system and participates in the delocalized molecular orbital network (Griffiths, Nassau b). The ground state bonding molecular orbitals have more electron density on the electronegative heteoratoms of the donor group. The electronic transition that produces color in these systems occurs when an electron is promoted from a pi orbital which has a high probability of finding the electron on the heteroatom of the auxochrome to a pi* orbital that has lower probability of finding the electron on the heteroatom. In quantum mechanical terms we can again say that electron density was transferred from the heteroatom of the auxochrome to another part of the molecule called the acceptor.

There are two types of acceptors, simple acceptors and complex acceptors. In a simple donor-acceptor complex the electron density is transferred to a small group of atoms on the molecule (Griffiths). In these systems it is possible to identify a few atoms that have increased electron density in the pi* orbital relative to the pi orbital. A simple acceptor group contains at least two multiply bonded atoms, the terminal atoms of which must be higher in electronegativity than carbon. These groups are often the chromophores identified earlier as being common in dyes. In these systems, the remainder of the molecule can be regarded simply as a conjugated bridge connecting donor and acceptor (Griffiths).

In the case of complex acceptors, it is not possible to identify any small group of atoms with increased electron density in the excited state (Griffiths). In these systems the electron density donated by the auxochrome is shared by all or most of the atoms of the molecule. The entire molecule serves as the acceptor. Most of the natural dyes in this book are complex donor-acceptor dyes. Napthaquinones such as juglone from walnut and lawsone from henna are often used as examples of donor-acceptor compounds as are anthraquinone dyes like alizarin from madder. The indigoid dyes are also complex donor-acceptor compounds (Griffiths).

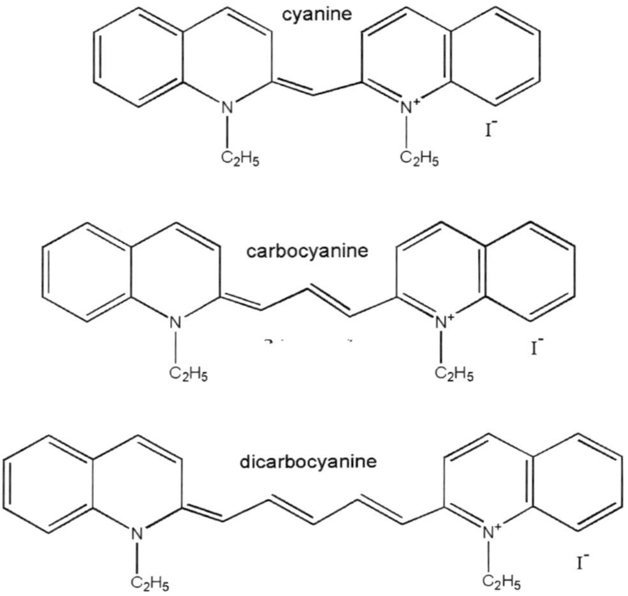

cyanine

carbocyanine

dicarbocyanine

dye	λ_{max}	λ range	dye color
1,1'-diethyl-2,2'-cyanine iodide	522 nm	~450-550 nm	red
1,1'-diethyl-2,2'-carbocyanine iodide	603 nm	~530-630 nm	blue-violet
1,1'-diethyl-2,2'-dicarbocyanine iodide	706 nm	~620-720 nm	cyan (B-G)

Figure 6.10 The cyanine dyes are used to model the simple quantum mechanical system "particle in a box" where the electrons are considered to belong to the entire molecule. Energy levels depend on the length of the box which changes with the length of the conjugated carbon chain. Different lengths produce different absorption maxima and thus differently colored dyes. This exact problem is a typical laboratory experiment in physical chemistry courses.

It is also possible to model an entire dye molecule as a quantum mechanical system without reference to chromophores or auxochromes. The simplest quantum mechanical system is a particle constrained to remain inside a box by walls of infinite energy, the so-called "particle in a box" problem (Shoemaker, Sime). Surprisingly this very simple system can be applied to the cyanine synthetic dyes and yields reasonable results. In this model, the electrons are the particles and the "box" is the conjugated pi system of the dye. Increasing the length of the methylene chain increases the length of the box and affects the energy of the orbitals. This changes the color of the dye as shown in fig. 6.10. Notice that the longer methylene chain reflects a blue-green color. This is expected if the energy levels are closer together, the absorptions are lower in energy and thus the red end of the visible spectrum is absorbed. The observed colors of the dye series follow the order predicted in table 6.1 with the intermediate dye appearing violet and the shortest dye of the series appearing red.

The advent of modern computers made the distinction between sigma and pi unnecessary without resorting to over-simplified models like the "particle in a box." In these computational methods many different atomic orbitals from each atom can be combined to produce a set of molecular orbitals. At this level it is no longer possible to say whether an orbital is sigma or pi or which atomic orbitals compromise a "bond." Molecular orbitals are given the symbol psi or phi and are numbered consecutively from lowest to highest energy. Some of the computational quantum mechanical methods use experimentally determined empirical parameters to simplify the computations and are called semi-empirical methods. Other more sophisticated methods called *ab initio* use no empirical parameters at all, but require significantly more computer time.

These methods can be applied to dyes as well; however modeling large molecules with *ab initio* methods requires a large amount of computer time (Griffiths). Semi-empirical quantum mechanical methods are sometimes used to study dyes (Christie). Often non-quantum mechanical or empirical methods are used as are qualitative discussions of electron density and sigma and pi molecular orbitals. More sophisticated methods may yield more accurate calculations of transitions energies but the basic result is still the same. In any method, the electronic transitions that produce color are usually the HOMO-LUMO transitions or those from the Highest Occupied Molecular Orbital to the Lowest Unoccupied Molecular Orbital (Christie, Griffiths).

We began this section on the chemistry of dyes by asking what type of molecule absorbed light in the visible region of the spectrum. In summary the answer is a conjugated molecule with resonance and delocalization of electrons. It is often large and aromatic. It must have either a very long polyene chain or one or more electron donating (electron rich) groups and one or more electron accepting (electron poor) groups (Swain 1976, Trotman). The more electronic transitions that are possible in a highly conjugated molecule, the closer they will be in energy (Trotman). Alternately, electron donating functional groups, or auxochromes coupled with suitable electron acceptors containing chromophores will behave as donor-acceptor complexes and have molecular orbitals with transitions in the visible range. Any molecule that absorbs light in the visible wavelengths will be colored.

Once dye chemists understood the way conjugation, auxochromes and chromophores worked, they were able to modify known dyes to change the colors in a rational way rather than by trial and error. Today they can use computer modeling to screen potential dyes for good candidates without having to synthesize each and every possibility. Now dye chemists can link separate dyes by a conjugated chain to produce a third dye of an unrelated color. Or they can connect two dyes with a saturated chain so the two pi systems cannot interact in essence producing two dyes that can be applied simultaneously. The resulting dye will absorb as both dyes individually. This procedure has given a solution to the elusive green dye problem. A blue dye and a yellow dye can be tethered by a saturated chain to produce a green dye (Trotman). Gone are the days of separately dyeing a fabric first with indigo and then with a yellow dye to produce green fabric. Gone as well are the problems of the less wash-fast dye washing out more quickly rendering the fabric yellow or blue rather than green. When the dyes are covalently linked, they will both have identical wash-fastness. The garment may fade with washing, but it will remain green. Quantum mechanics and modern chemistry have revolutionized the world of dyes.

Dye Chemicals Featured in this Book

All the dye chemicals featured in this book exhibit the features of conjugation, resonance and delocalization described in this chapter. Most contain chromophore functional groups as well as hydroxyl groups, amines or other auxochromes. Anthocyanins and anthochlors can be used to demonstrate the pH dependence of color. Indigo dyeing is dependent upon redox reactions and the concepts of chemical oxidation and reduction can be taught and observed in lab using indigo. Applications of some of these concepts are included in the prepared activities in part four.

Often many different dye plants contain similar or identical dye chemicals. Some of these ubiquitous and historically important dye chemicals include luteolin, indigo, quercetin, quercetagetin, iron oxide, juglone and iron-tannate complexes presented in earlier chapters. In most cases, the chemistry of these dyes has been well studied and they serve as good examples for the general structures of natural dyes. By virtue of their occurrence in a number of different dye plants, they are also good examples to use with students. We will end our discussion of the chemistry of dyes for now by looking at the structures of these dye chemicals. Further information about the biochemistry and chemical reactions of the dyes is presented in Ch. 8.

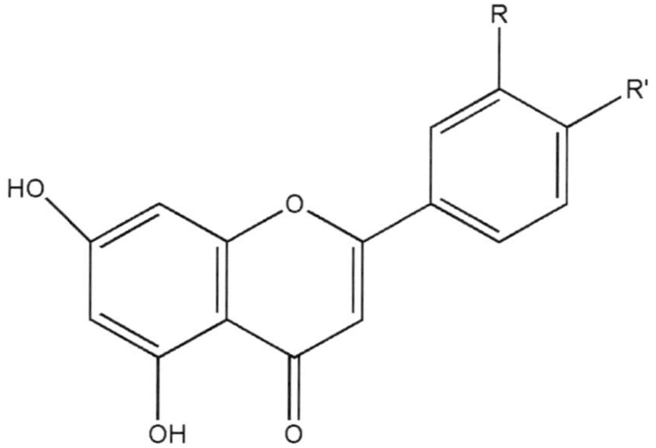

	R	R'
a	OH	OH
b	H	OH
c	H	H

Figure 6.11 The structures of some important flavone dye chemicals. a) Luteolin and b) Apigenin are found in chamomile and weld plants. c) The general flavone core structure.

Luteolin (fig. 6.11a) and apigenin (fig. 6.11b) are chemicals known as flavones (fig. 6.11c). Flavones are very commonly found in plants. Luteolin and apignein are the major chemical species responsible for the dyeing properties of weld and chamomile plants (Cannon & Cannon). Both plants produce a yellow dye. Weld is one of the oldest dyestuffs found in

Europe. Chamomile is frequently used as a dyestuff by modern hobbyists. Additional information about both dyes and recipes for their use can be found in Ch 5.

Quercetin (6.12a) is a ubiquitous chemical and is found in a number of dye plants (Mayer & Cook, Swain 1980), including black oak bark, onion (Cannon & Cannon), *Euphorbia* species (Gallagher) and at least two species of eucalyptus (Glasson & Glasson, Cribb & Cribb). Quercetin is a flavonol and is described in more detail in Chapter 3. Quercetin alone produces a yellow color. The use of this dye from black oak was very important in the early American dye industry (Butterworth).

Quercetagetin (fig. 6.12b) is the second important dye chemical found in black oak and marigold species (Cannon & Cannon). It is a historically important dye chemical in pre-Colombian America and Africa (Colton, Harris). Cannon & Cannon report that quercetagetin produces a more orangish color than quercetin. Quercetagetin is less common in plants than quercetin which is widely distributed (Harborne 1976). It is described in more detail in chapter 3.

Quercetin and quercetagetin are both flavonols. The related compound morin (fig. 6.12c) is found in the wood of osage orange and mulberry. Additional information and recipe for the use of osage orange can be found in Ch. 4. Other flavonols discussed in this book are kaempferol (fig. 6.12d) found in goldenrod, fisetin (fig. 6.12e) found in sumac and patuletin (fig 6.12f) found in chamomile. Goldenrod and sumac are discussed in Chap.3 and chamomile in Ch. 5. Many other dye plants contain these or related flavonols including tea (Cannon & Cannon) and flavonols are widely distributed in nature (Goodwin & Mercer, Harborne, 1967 and 1976). The core flavonol structure is shown in fig 6.12 g.

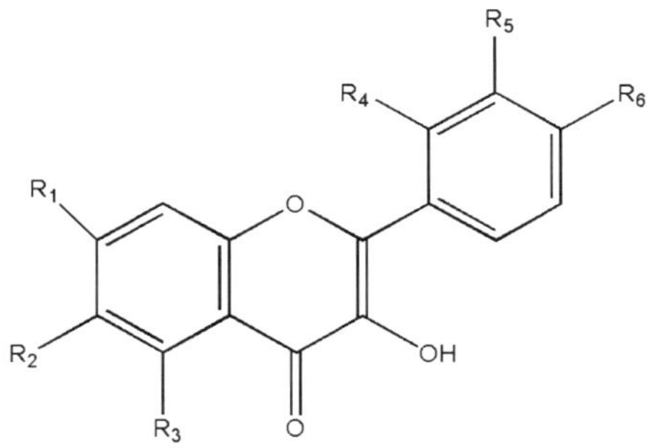

	R_1	R_2	R_3	R_4	R_5	R_6
a)	OH	H	OH	H	OH	OH
b)	OH	OH	OH	H	OH	OH
c)	OH	H	OH	H	H	OH
d)	OH	H	OH	OH	H	OH
e)	OH	H	H	H	OH	OH
f)	OH	OCH_3	OH	H	OH	OH
g)	H	H	H	H	H	H

Figure 6.12 Flavonols are common dye chemicals. a) quercetin found in oak, onion and eucalyptus b) quercetagetin found in oak and marigold c) morin found in the wood of osage orange and mulberry d) kaempferol found in goldenrod e) fisetin found in sumac heart wood and yellow cedar f) patuletin found in chamomile. (Chamomile also contains luteolin and apigenin shown in fig. 6.11). g) The flavonol core structure.

A class of flavonoid dyestuffs frequently used to produce reds and blues are the anthocyanins. Anthocyanins are found in hibiscus, rose and cranberry used as dyes in this book as well as most red and blue flowers and fruits. The color of anthocyanins is pH dependent. In general, acid pH's favor the red colors and alkaline pH's favor the blue colors. This interesting chemistry is described in Ch. 8. The pH dependent color change is the basis for a student experiment in Part 4. The general structure of anthocyanins as well as some specific molecules are shown in fig. 6.13.

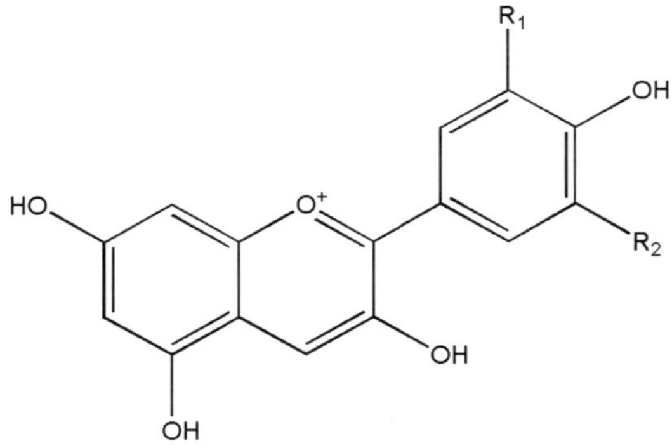

	R_1	R_2
pelargonidin	H	H
cyanidin	OH	H
delphinidin	OH	OH
peonidin	H	OCH_3
petunidin	OCH_3	OH
malvidin	OCH_3	OCH_3
anthocyanidin core	H	H

Fig. 6.13 Anthocyanins are a class of dye chemicals that produce red and blue colors. (Cutright et. al, Harborne 1976 and 1994)

The quinoids, like the flavonoids are a large category of natural products including natural dyes. Juglone (fig. 6.14a) is responsible for the dye properties of walnut species. The ablility of walnut to serve as a dye was known at least since the Roman Empire and was commonplace in colonial America. The most abundant source of the dye is the green nut hulls of the American Black Walnut tree. The details of this dye are found in Chapter 4. It is closely related to the dye chemical lawsone found in Henna (fig. 6.14b). Both dyes are naphthaquinones (Fig. 6.14c). The related anthraquinone (fig. 6.14d) dyes are an important class of natural dyes found in about half of all flowering plants but chiefly those from six families (Thomson). They include many well known historical dyes such as alizarin from madder (fig. 6.14e) and the closely related red insect pigments including cochineal, lac and kermes. The dyes may be present in all parts of the plant, but are generally found only in wood and root (Thomson). They have not been featured in this book because the better known anthraquinone dyes are mainly from tropical plants, however no discussion of quinoid dyes would be complete without mentioning their importance. The interested reader can find information about madder and other anthraquinones in a number of the references.

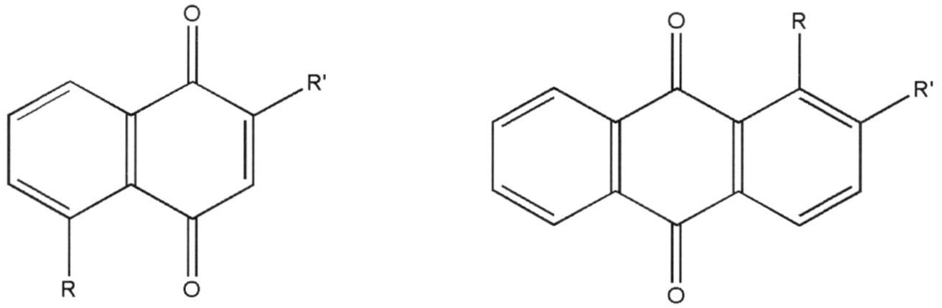

	R	R'
a	OH	H
b	H	OH
c	H	H

	R	R'
d	H	H
e	OH	OH

Fig. 6.14 The quinoids are a large category of natural coloring matter. a) Juglone is the naphthoquinone dye chemical in walnut species. b) Lawsone, the red dye in Henna is closely related to juglone. c) The naphthoquinone core structure. d) Anthraquinones are related to naphthoquinones. The anthraquinone core structure has one additional ring. e) Alizarin from madder roots, one of the most well known natural dyes, is an anthraquinone.

Iron Oxide may well be one of the earliest dyestuffs and one of the most universal. Its use began with primitive man and persisted up to the current times. Iron oxide is an unusual natural dye because it is an inorganic chemical or mineral dye rather than an organic chemical or vegetable dye. The same chemical species, iron (III) oxide, is responsible for rust and may be considered a stain more than a dye by some (Liles). Iron oxide is also a useful mordant, as described earlier. Historical information and recipes for iron oxide dye can be found in Chapter 3.

Iron-tannate complexes have also been used by many cultures in many different recipes. The dye stuff is a very large complex of iron metal with naturally occurring tannins. The general structure of the simplest hydrolyzable tannin is shown in Fig. 6.15. The production of iron-tannate complexes can be similar to indigo in requiring precise and careful chemistry to achieve the correct results. Like indigo, the production of iron-tannate dyes had superstitions and traditions associated with the dyeing process in many cultures. The production of iron-tannate dyes is described in Chapter 3 along with a North American recipe. Tannins themselves are frequently used as mordants with cotton. Tannins are important components of many substantive dyes including black walnut, oak, coffee, tea and mint. Tannins are discussed further in Ch. 7.

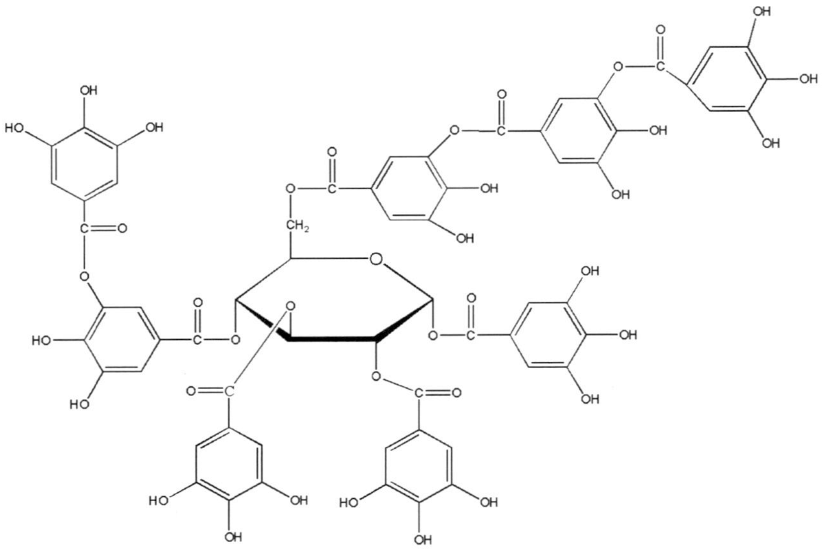

Fig. 6.15 Tannins are a class of related compounds of variable structure. The general structure of "Chinese tannin" the simplest hydrolysable tannin is shown here.

Although flavonoids and quinoids make up a large number of natural dyes, they are not the only important dyes. Indoxyl (fig. 6.16a) is the precursor to the dye chemical indigotin (fig. 6.16b) of the famous blue dye, indigo. Indoxyl and closely related chemical species are found in a large number of plants of varying genus and species throughout the world. Indigotin is perhaps one of the most well known and universal dye chemicals even though the process of producing indigotin from indoxyl and similar precursors is quite complex. It can be used as an example of oxidation-reduction processes in the chemistry classroom. It can also serve as a lesson that so-called "primitive cultures" were capable of quite complex chemistry in the production of dyes. Although their dye practices were ruled by tradition and laced with superstitions well into the Middle Ages, their ability to reproducibly follow a complex dye procedure should not be minimized. In fact, many chemistry students would be happy to see such good reproducibility in their own results! The process of chemical conversion of indoxyl to indigotin is explained in Chapter 3 and shown in fig 3.1.

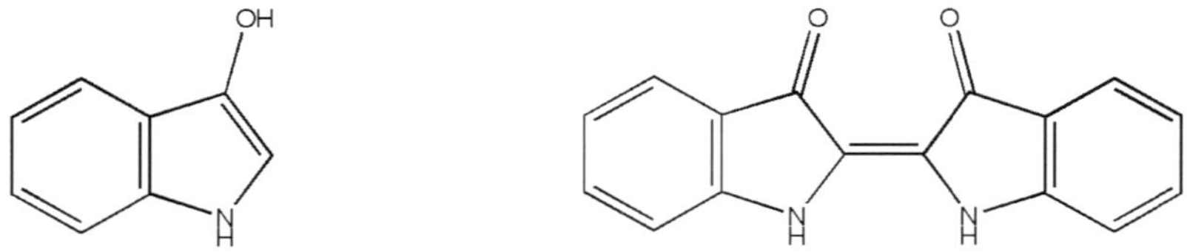

a b

Fig. 6.16 The monomeric and dimeric species involved in indigo dyeing. a) Indoxyl is extracted from indigo plants. b) Indigotin is produced as a result of the vat fermentation process. The chemistry of indigo is described further in Ch. 3.

Closely related to the flavones and flavonols are the anthochlor pigments. Like the anthocyanins, the anthochlors undergo pH dependent color changes as discussed in Ch. 8. Anthochlors are important dye chemicals in some composites including coreopsis, dahlia and cosmos. Chalcones (fig. 6.17a) including butein (fig. 6.17b) marein (Fig. 6.17c) and stillopsin (fig. 6.17d) are found as glycosides in coreopsis species (Boehm). Closely related chalcone glycosides are found in dahlia and cosmos. Aurones (Fig 6.17e) including sulphuretin (Fig. 6.17f), marimetin (Fig. 6.17g) and leptosin (fig. 6.17h) and related glycosides are found in these flowers. Different glycosides often have different common names and sometimes the same structure can be found with slightly different names in different references. The similarity of the chalcone and aurone common names also makes referring to these compounds difficult.

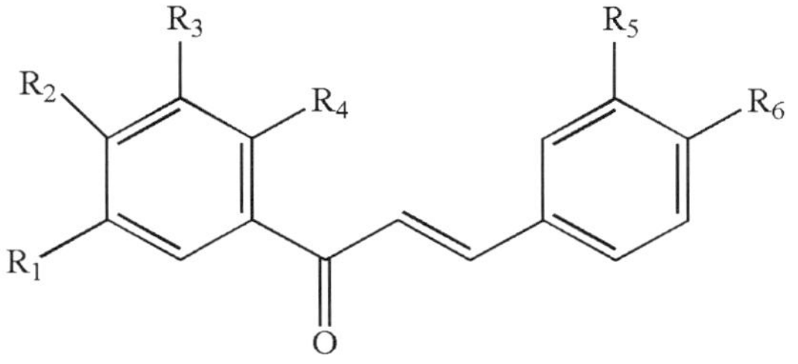

R_3 R_5 R_2 R_4 R_6 R_1 O

chalcone

	R_1	R_2	R_3	R_4	R_5	R_6
a	H	H	H	H	H	H
b	H	OH	H	OH	OH	OH
c	H	OH	OH	OH	H	OH
d	OH	OH	H	OH	OH	OH

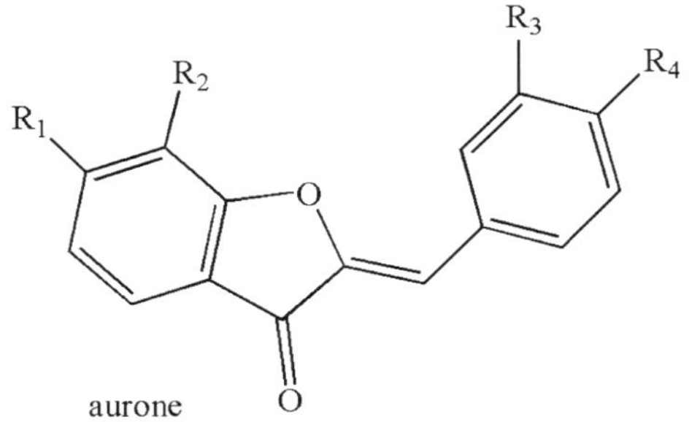

R_3 R_2 R_4 R_1 O O

aurone

	R_1	R_2	R_3	R_4
e	H	H	H	H
f	OH	H	OH	OH
g	OH	OH	OH	OH
h	OH	OCH_3	OH	OH

Fig. 6.17 Dye chemicals called anthochlors are found in some composites. a) The core structure of the common chalcones. Chalcone pigments include b) butein c) marein and d) stillopsin. e) The core structure of the rarer aurone pigments. Aurones include f) sulphuretin, g) marimetin and h) leptosin.

Other unusual flavonoid pigments include the isoflavones. Isoflavones are only found in higher plants. Two isoflavone pigments, osajin (fig. 6.18 a) and pomiferin (fig. 6.18b) are found in the fruit of the osage orange tree. The wood contains morin, a flavonol.

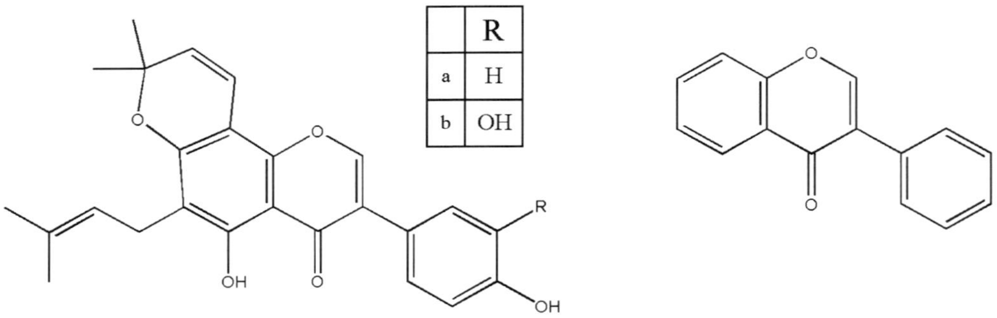

	R
a	H
b	OH

a) and b)

c) isoflavone core

Fig 6.18 The dye chemicals found in osage orange fruits are isoflavones. a) Osajin and b) pomiferin. Morin (fig 6.12) found in the wood of the tree is not an isoflavone. c) The general isoflavone core structure.

The alkaloid pigment berberine, from the barberry plant (fig. 6.19), is used as a dyestuff in this book. Other alkaloids dyes are known but are not used in this book. Alkaloids are not as common among natural dyes as flavonoids and quinoids. Berberine is also the only natural dye that is in the category of basic dyes (Colour Index).

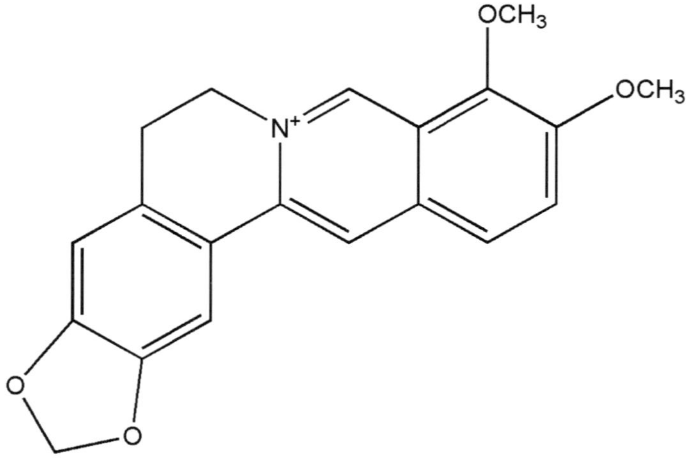

Fig 6.19 Berberine from barberry is an alkaloid and is structurally different from the more common plant dyes. It is the only natural dye to be classified as a basic dye.

Carotenoids are important plant pigments that are long conjugated hydrocarbon chains rather than cyclic aromatic molecules. Although no carotenoid dyes are specifically used in this book, they need to be included in any discussion of plant pigments and dye chemicals. The most well known carotenoid dyes are carotene (fig 6.20a) from carrots and other orange vegetables, and lycopene (fig. 6.20b) a redder pigment from tomato. Historically important carotenoid dyes include annatto or bixin sometimes used as a food dye and the yellow spice saffron (Cannon & Cannon).

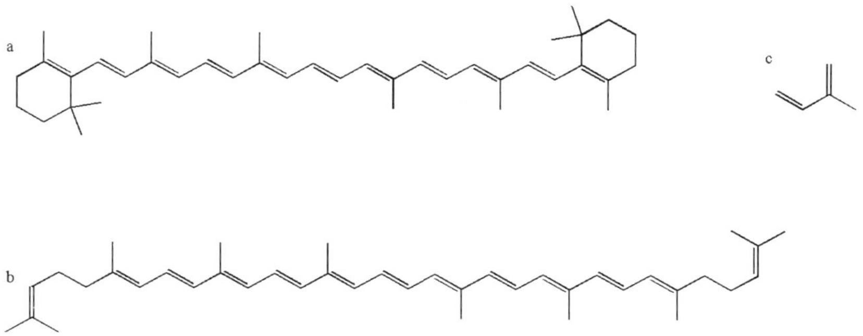

Fig.6.20 Carotenoid pigments a) carotene and b) lycopene c) the building block of carotenoids, the "isoprene unit"

Oxidation of hydroxyflavans (fig 6.21) produces a brown color in chocolate, tea and browning fruit (Swain 1976). It contributes to the color of tea dye and possibly other dyes as well.

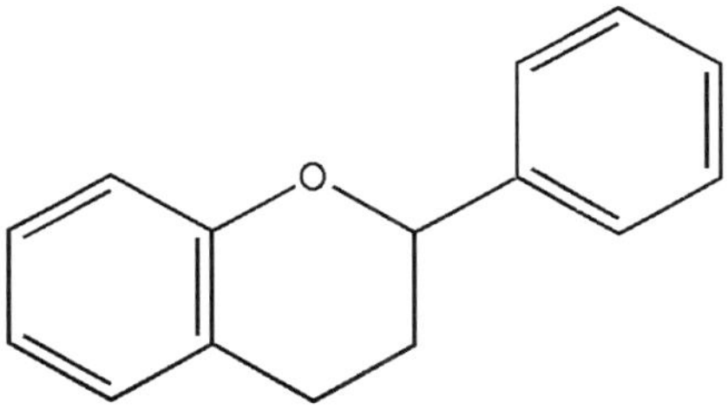

Fig 6.21 Core flavan structure. Oxidation of hydroxyflavans is responsible for the browning of fruit and the brown color in tea and chocolate.

Chapter 7
Fibers, Fabrics and the Binding of Dyes

The physics and chemistry of dyes and dyeing was not understood for most of the history of dyes. Dye practitioners knew that certain things worked by trial and error. Knowing what sorts of things worked allowed them to deduce others. In this way additional dye sources were tried and various additives tested to render a dye fast. The chemistry of dyes binding to fabrics is not simple and even today is not completely understood. The focus of dye chemistry research is on the synthetic dyes with considerably less research using natural dyes. Fortunately for us, many of the results of research conducted using synthetic dyes can be applied to an understanding of natural dyes as well.

Fibers and Fabrics

The process of dyeing begins with a fiber or fabric to be dyed. The types of dye that will be fast, and how a dye binds to the fiber depends on the nature of the fiber. Fibers can be divided into categories by different classification systems. The fibers and fabrics here will be limited to natural fibers. Dyers traditionally divide natural fibers into animal fibers and vegetable fibers depending on the source of the fiber (Androsko, Gohl & Vilensky, Storey, Thompson & Thompson, Trotman, Wingate). Chemical classification of natural fibers divides them into protein fibers and cellulosic fibers depending on their chemical make-up. Luckily, the categories are identical in both systems. All animal fibers are protein fibers and all vegetable fibers are cellulosic fibers (Trotman, Storey, Thompson & Thompson). A simple classification system is shown in Table. 7.1.

Table 7.1 Natural fibers classified as protein or cellulosic in nature. Each broad category can be subdivided still further. Other protein and cellulosic materials have been dyed with natural dyes throughout history and some are included here.

Fibers and Fabrics

Protein: Keratin	Protein: Fibroin	Protein: Collagen	Cellulosic
Wool	Bombyx silk	Leather	Cotton
Cashmere	Tussah silk	Animal hides	Flax
Angora			Linen
Alpaca			Jute
Rabbit hair			Ramie
Mohair			Sisal
Other Materials			
Quills			Raffia
Feathers			Wood
Fur			Grass
			Reed
			Cane/rattan

All protein fibers share the common features of proteins, that is, they are polymers of amino acids formed by the condensation of amine and carboxyl groups to produce amide bonds. The exact arrangement of the amino acids in a protein however, results in very different structures and functions for the thousands of proteins in nature. The proteins found in animal fibers belong to the category called fibrous proteins. Fibrous proteins have structural roles in animal tissues being components of bone, cartilage, ligament and tendon, skin, blood vessels, arteries, hair, nails and claws, insect 'silk' secretions and so forth. The amino acid composition and arrangement in fibrous proteins is very variable depending upon their intended function. Similarly the arrangement of the same proteins within various cells or types of tissue also varies according to the function of that tissue.

Some authors subdivide protein fibers into wool, other animal hairs, and silk (Thompson & Thompson). Non-wool animal hairs in their system include mohair, alpaca cashmere, rabbit, however from a chemical viewpoint, wool and other animal hairs are similar. All hair and fur is primarily composed of the helical protein keratin shown in fig 7.1. Some of the keratin is arranged in an orderly manner in the hair cells in small groups called fibrils and the fibrils are arranged in an orderly manner within hair fibers. This orderly crystalline region is between 25%-30% in wool (Gohl & Vilensky). The remaining 70%-75% of the keratin in wool is in disordered or amorphous regions of the fiber (Gohl & Vilensky).

The outside of the wool strands is called the cuticle and is composed of overlapping surface cells or scales (Gohl & Vilensky, Thompson & Thompson, Trotman). This is covered by a waxy, water repellant coating called the epicuticule. Although all hairs are made of keratin, the microscopic structure of the hair varies considerably in different species. Most non-sheep hairs have smaller scales than wool or no scales at all (Wingate). These differences in cuticle structure are most likely responsible for differences in the degree of uptake of dyes by different types of animal hair. Some differences may also be due to the arrangement of the keratin fibrils within the animal fiber and by the cellular structure of the hair. Dye binds only to the amorphous regions of the fiber because the dye molecules cannot fit into the tightly packed crystalline regions (Gohl & Vilensky). Differences in the ratio of crystalline to amorphous regions will affect dye binding.

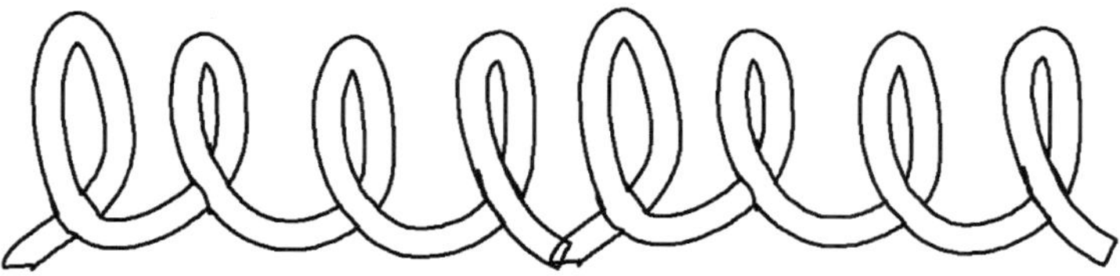

Fig. 7.1 Fibrous proteins are polymers of amino acids linked by amide bonds. The side chain of the protein may be nonpolar, neutral polar or charged depending on the amino acids. A schematic diagram of keratin, found in wool and hair is shown here.

Keratins are held together in the hair fibrils by covalent disulfide bonds between amino acids of adjacent keratin molecules and by interactions of oppositely charged groups called salt bridges or ion pairs. The helical conformation of keratin and the arrangement of the disulfide bonds accounts for the ability of wool to stretch under tensile force and then to return to its original size and shape. The arrangement of the disulfide bonds also accounts for differences among hair of different individuals, such as straight or curly, but is not known to play any role in the dyeing process. The charged groups on the proteins that comprise salt bridges however are very important in the binding of dyes as will be discussed later in this chapter.

Silk while also a fibrous protein, is chemically very different from wool. Silk is the secretion of a silk worm and is harvested from the cocoons. Reportedly the best silk is produced by *Bombyx mori* silk worms cultivated in China and Japan but other types of silk worms produce silk as well. These "wild silk" varieties are coarser than the *Bombxy* silk and are known as "tussah," "shantung" and "honan" silks (Storey, Thompson & Thompson). A schematic diagram of silk protein, called fibroin, is shown in fig. 7.2. Fibroin has fewer charged amino acid side chain than keratin and has an extended conformation rather than a helical shape of keratin.

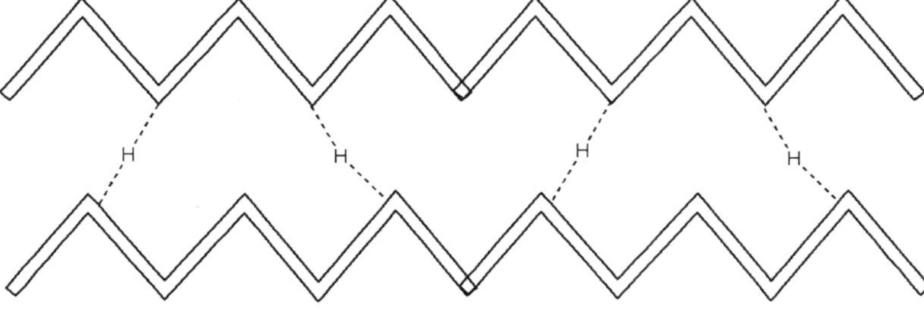

Fig. 7.2 Fibroin is a fibrous protein found in silk. A schematic of fibroin is shown here.

The fibroin chains are arranged into sheets rather than coils like keratin. The fibroin molecules in the sheets are held together primarily by hydrogen bonding between the adjacent protein backbones. The stacked sheets of protein are attracted by van der Waals forces. This different protein conformation of fibroin accounts for silk's physical properties, such as its great tensile strength and its inability to stretch as compared to wool. These same differences in the protein confer silk with different chemical properties. Consequently, even though it is a protein fiber, silk dyes differently from wool. With natural dyes, silk must be mordanted more like cotton but takes dye well (Liles). This is most likely due to the shortage of charged amino acid side chains so important in dyeing of wool. Interestingly, silk can be dyed with synthetic dyes intended for both wool and cotton (Thompson & Thompson).

Porcupine quills, leather and other animal materials are not included in Thompson & Thompson's system although they are also protein in nature and can be dyed. Some Native American tribes decorated clothing with dyed quills. Quills like other nails and claws are mostly keratin as are hair, fur and feathers. The arrangement of keratin molecules within a quill or nail and the amount of cross-linking between the keratin molecules confer quills with quite different structural properties from hair. The same dyes that dye wool are applicable to all keratin materials although the different molecular arrangements would probably require some differences in processing and differences in the affinity toward dyes. Hides are primarily composed of the protein collagen and should have some affinity for protein dyes. Liles has dyed leather with indigo (Liles) and many traditional leather tanning processes use the same chemicals used as dyestuffs (Mead, Welch & Coombs). Carman has dyed feathers with eucalypt dyes (Carman, 1984) and Buchanan refers to dyeing of leather, feathers and fur with natural dyes as well (Buchanan).

Like the protein fibers, the vegetable or cellulosic fibers can be subdivided. The classification includes cotton, harvested from the seeds of *Gossypium sp.* (Storey, Thompson & Thompson) the bast fibers from the stem parts of plants which include flax and linen, jute, ramie and hemp, and finally leaf fibers of sisal (Ponting, Thompson & Thompson) and grasses. All of these fibers are similar chemically being composed primarily of cellulose, a linear polymer of glucose. The polymer is typically made from about 10,000 glucose monomers (Gohl & Vilensky). A schematic of cellulose is shown in fig. 7.3. The chemistry of cellulose is important for dye binding, particularly with synthetic "reactive dyes."

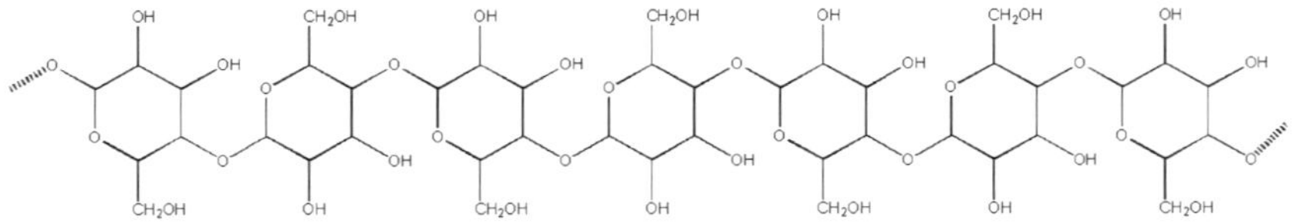

Fig. 7.3 Cellulose is a linear polymer of glucose. The individual glucose monomers are bonded by beta 1,4 glycosidic bonds.

With both synthetic and natural dyes, the microscopic arrangement of cellulose in the fibers is also important for dye binding. A cotton fiber is a single plant cell best described as a twisted ribbon or collapsed tube (Gohl & Vilensky, Storey). Cotton fibers have a cuticle but unlike wool the cotton cuticle is a waxy layer a few molecules thick making it more like the epicuticle of wool. The boiling and bleaching procedures of cotton processing remove most of the cuticle (Gohl & Vilensky, Storey). The resulting cotton is nearly pure cellulose (Christie, Storey). The cellulose polymers in the cotton cell are held together by intermolecular hydrogen bonds between hydroxyl groups. Cellulose in cotton fibers is 65%-70% crystalline and 30%-35% amorphous (Gohl & Vilensky). The open structure of cellulose allows large dye molecules to enter the fiber (Christie). Dyes bind only to the amorphous regions of the fiber, however, because they cannot fit into the more tightly packed crystalline regions (Gohl & Vilensky). Fibers from different plants will vary in cell size and structure. The cellulose will vary in length and the ratio of crystalline to amorphous regions. Because of these differences, different cellulosic fibers will take up dye differently and with varying fastness.

Further evidence of the importance of the arrangement of the cellulose within the fibers comes from the processing of cotton. Commercial cotton is usually treated by a process called Mercerization. Mercerization is done by first placing the cotton fabric under tension to prevent shrinkage and then treating it with a sodium hydroxide solution (Storey, Thompson & Thompson, Trotman). Mercerization of cotton causes a change in the arrangement of the cellulose molecules within the fiber (Trotman). Mercerized cotton has a more lustrous appearance and better drape than untreated cotton as well as increased affinity for dyes (Storey, Thompson & Thompson, Trotman). Processing cotton also produces carboxyl groups in cotton (Motomura *et. al.*).

Most basket materials are bast or leaf and some traditional peoples in American, Africa and Australia dyed basket materials. In America the Tlingit of Alaska used roots and berries as dyes (Wright) and the Appalachian peoples used walnut dye (Cincinnati Nature Center). In Africa the Kikuyu of Kenya used bark to dye baskets (Wright) and Australian Aboriginals of Queensland used berries and leaves to dye twined baskets called dilly bags (Lloyd). Some bags and baskets are painted with ocher rather than dyed (Were, Wright). A number of current aboriginal artists of the rainforest peoples make traditional baskets from pandanus palm fibers or watul grass and dye them with natural dyes obtained from the bark of various roots (Were). Australian hobbyist Jean Carman reports that sisal, corn husks and other basketry materials are dyed well with eucalypt dyes (Carman, 1984).

Basket materials are not commonly included in dye manuals and only a few basket making books include recipes for natural dyes (Navajo School). Basket materials are primarily cellulose and can be dyed with natural dyes used for cotton and other cellulosic materials. Some authors report that the dyes are not as fast with these materials (Wright) but this is probably due the method of dyeing rather than the nature of the materials. Cotton is traditionally dyed at simmering temperatures whereas many baskets are dyed at room temperature (Wright, Cincinnati Nature Center). The Navajo recommend simmering dye baths when dyeing raffia for baskets and they do not report problems with fastness of the dyes, but rather prefer them for "lasting color" (Navajo School). Similarly the rainforest Aboriginal artists boil the fibers in the dyebath (Were). Buchanan refers to the dyeing of wood, paper, reeds and stems for baskets with natural dyes (Buchanan).

The Theory of Dye Binding Fibers

There are three stages in the process of dyeing (Aspland, Trotman). The first of these stages is diffusion of the dye chemical from solution in the dyebath to the surface of the fiber and its adsorption onto the surface. This is sometimes called the strike. The second is diffusion of the dye into the fiber and involves some complex chemistry. The third is fixation or permanent binding of the dye to the fiber and is probably the slowest step. Each of these important stages is described in detail below.

The first and second stages of dyeing are very similar regardless of the fiber used. These processes involve quite complex chemistry including the physical chemistry of diffusion and various models of binding equilibrium. These could be used in the teaching of college Physical Chemistry courses. The salient points will be summarized here including some well known mathematical relationships. In all cases, the processes are described conceptually as well as mathematically for readers not interested in the mathematical rigor. The reader interested in more detail and more mathematical description is referred to Trotman.

In spite of a great deal of research there is still not a complete understanding of the binding of dyes to fibers, particularly with natural dyes and cotton. It is clear that the forces responsible for keeping a dye molecule fixed on wool and other protein fibers are different from those involved with cotton. These are important for the third stage of dyeing, binding the dye permanently to the fiber, and will be of interest to all teachers of chemistry. Teachers interested only in the chemistry of this stage of dye binding may choose to skip the more mathematical treatment of dye entry into the fiber.

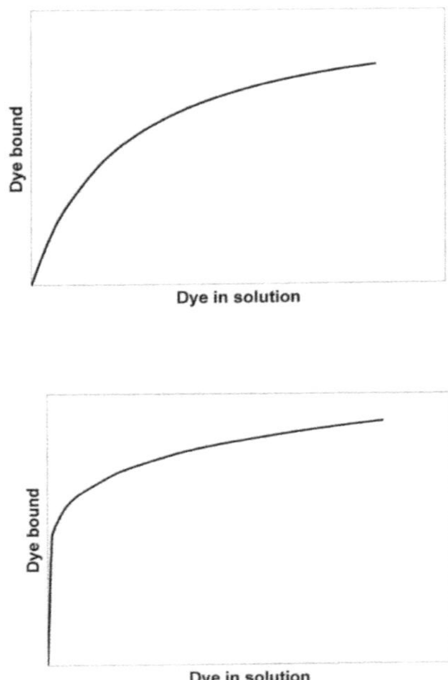

Fig. 7.4 Ideal curves for dye binding to fabric. a) Langmuir or saturation curve observed with protein fibers. b)The Freundlich curve is more descriptive of cellulosic fibers because they do not have specific binding sites for dye.

The most important factors in migration of the dye to the fiber are temperature, agitation and attractive forces (Aspland, Trotman). The first two underscore the importance of dyeing at elevated temperature and stirring the dye solution. Increased temperature causes increased movement of the molecules. Agitation will increase the rate of dyeing by bringing the dye to the surface of the fiber more quickly. The electrostatic attractions are sometimes controlled by maintaining proper pH. Often it is electrostatic forces that attract the dye to the surface of the fibers and hold them there. This results in an increased concentration of dye molecules in the vicinity of the fiber as compared to the dyebath and aids diffusion into the fiber itself, the second stage of dyeing (Trotman). In other cases dyes are not attracted electrostatically, but are held at the surface instead by a variety or forces including van der Waals forces, hydrogen bonding, dispersion forces and hydrophobic effects (Aspland).

The first stage of adsorption of dye onto the surface of the fiber is an equilibrium process. There are several mathematical models that describe the adsorption equilibria, and the best model depends on the type of fiber. The Langmuir equilibrium shown in fig. 7.4a describes the observed behavior of wool and other protein fibers or any anionic dyes attraction for oppositely charged fibers (Aspland, Trotman). This is a saturation curve, meaning that wool has a certain number of sites available to bind dye. When all the sites are filled, the wool is saturated with dye and no more can be bound regardless of the amount of dye available. This type of behavior corresponds to the known chemistry of wool dye complexes described later in this chapter.

Cellulosic fibers show a different type of adsorption behavior and are usually described by a Freundlich equilibrium (Aspland, McGregor, Trotman). A Freundlich curve is shown in fig 7.4b. The Freundlich curve is different from a Langmuir curve because strictly speaking it does not saturate or become horizontal at higher dye concentrations. This type of adsorption is seen when there are no specific binding sites for the dye, so no sites to saturate (Aspland, Trotman). The curve begins to level off at higher concentrations because all the easily accessible places for dye to bind are occupied (Trotman). This fits with the known chemistry of cellulose described later in this chapter and indicates non-ionic, or weak bonding forces between fiber and dye (Aspland).

From a pragmatic point of view, the saturation behavior of protein fibers and the near saturation behavior of cellulosic fibers explain why it is possible to add more water to a dyebath without producing a paler color on the fiber. The recommended ratios of dyestuff to fiber weights are intended to ensure that the concentration of dye chemical the dyebath is in the region where the Langmuir curve is horizontal and the Freundlich nearly so. At these points changes in concentration of the dyebath have little effect on the amount of dye bound to fiber. Conversely crowding the dyebath with fiber will result in uneven dyeing.

This saturation behavior also explains the technique of exhaustive dyeing. In exhaustive dyeing subsequent batches of fiber are dyed after an initial dyeing. The subsequent batches become increasingly paler. At this point, the dyebath concentration is in the region of the curves where the slope is steep. Changes in the concentration of the dyebath do affect the amount of dye bound to fiber. Each batch begins at a lower dyebath concentration and thus a smaller amount of dye will enter the fiber.

Stage two, or the entry of the dye chemical from the surface into the fiber is the most rapid stage of dyeing. It is so rapid relative to diffusion in the dyebath that it will quickly reduce the local concentration of dye at the surface of the fiber if the dyebath is not agitated (Trotman). It is also the most complex and mathematically detailed of the three stages. Readers not interested in the mathematical complexity may skim this section to understand the relevant concepts or even skip it entirely.

The driving force for this process is the higher chemical potential of the dye in the aqueous versus the fiber phase (Aspland, Trotman). Salts in solution raise the chemical potential of the dye in the aqueous phase even further (Aspland). In other words, the process is driven by a decrease in Gibbs free energy of the system when the dye enters the fiber. The delta G for the process is negative or favorable because of both favorable enthalpy and entropy (Trotman). The relationship between free energy, enthalpy and entropy is given by eq 7.3 where delta G is the Gibbs Free Energy of the system,delta H is the enthalpy of the system, T is the temperature in Kelvin and delta S is the entropy of the system.

$$\Delta G = \Delta H - T\Delta S \qquad \text{Eq. 7.3}$$

The process of dye entering a fiber is exothermic, a favorable enthalpy change, thus enthalpy favors dyeing. The removal of the largely nonpolar hydrocarbon dye molecules from aqueous solution corresponds to an increase in entropy or disorder of the water (Trotman). This is often called "the hydrophobic effect." Water molecules form hydrogen bonded cages or clathrates around hydrophobic molecules and this corresponds to an ordered or low entropy situation. Removal of these nonpolar residues from the water results in less structure and more entropy in the aqueous phase. In the case of dye entering fiber, the increased entropy of the water phase more than makes up for the decreased entropy of the dye molecules crowded into the fiber and results in a favorable, positive entropy change for the process. Together the favorable enthalpy and entropy of the system result in an exergonic process, a favorable free energy change.

The process of dye entering fiber is usually described as diffusion toward the center of the fiber (Aspland, Trotman). In general, diffusion processes at constant temperature follows Fick's Law (eq. 7.4) where the ds/dt is the rate of diffusion, dc/dx is a concentration gradient and D is the temperature dependent diffusion coefficient, a proportionality constant (Trotman).

$$ds/dt = -D(dc/dt) \qquad \text{eq 7.4}$$

Conceptually, Fick's Law means that the rate of diffusion is proportional to the concentration gradient of the dye. In other words, diffusion is faster when the concentration of dye in the fiber is lower so dye tends to move into an undyed fiber very rapidly. In reality with a cylindrical shaped fiber the process requires somewhat more complicated mathematical treatment than Fick's Law, however the process is similar (Trotman).

The entry of dye into the fiber is temperature dependent both because the dye molecules must overcome an energy barrier to enter the fiber and because higher temperatures swell the fibers more increasing pore size (Gohl & Vinlensky). The energy barrier can be described as an energy of activation of the dye molecule (Gohl & Vilensky, Trotman) and as such follows a temperature dependent Boltzman-type distribution (Trotman). The temperature dependence is included in the diffusion coefficient, D as shown in eq 7.5 where Do and Dt are the diffusion coefficients at temperatures 0K and t respectively, Ea is the activation energy T is the temperature in Kelvin and R is the gas constant.

$$D_t = D_o \, e^{-(Ea/RT)} \qquad \text{eq 7.5}$$

The final stage of dyeing a fiber is the permament attachment of the dye molecules to the fiber molecules (Aspland, Trotman). This is the slowest of the three phases but the most important to ensure a fast dye (Trotman). Common causes of dye fixation are physical entrapment, a physical bond between the dye molecules and the fiber, the formation of a covalent chemical bond between dye and fiber and the conversion of a soluble dye chemical to an insoluble form (Aspland). The forces that fix a dye to a fiber vary with the dye and the fiber. In general different factors are important with protein dyes than with cellulosic dyes. The process of insoluble pigment formation is only important for vat dyes such as indigo.

In both cotton and wool, dyes are bound in the amorphous regions of the fibers because the tightly packed crystalline regions are too small for large dye molecules to fit. Physical entrapment is important for both these fibers. Heating the fiber during the dyeing processes swells the fiber and increases the size of the spaces between the polymers (Gohl & Vilensky). Dye molecules enter the larger spaces and upon cooling can be physically trapped in the fiber. This is particularly important with mordant and vat dyes (Gohl & Vilensky). Table 7.2 shows the size of typical mordant and vat dyes as compared to intermolecular distance between polymers in amorphous regions of both wet and dry fibers (Gohl & Vilensky). As can be seen from the table, large dye complexes can easily fit in the spaces in wet fibers but will be trapped within the dry fiber.

Table 7.2 Comparison of the sizes of dyes and the interpolymer spaces in fibers.

Dye	Size	Average Dimensions
1:1 mordant dyes	Average	1.2 nm x 0.8 nm x 0.4 nm
2:1 mordant dyes	Large and bulky	
Vat dyes	Very large	1.6 nm x 1.2 nm x 0.4 nm
Fiber	Interpolymer distance dry	Interpolymer distance wet
Cotton	0.5 nm	2-3 up to 10 nm
Wool	0.6 nm	Up to 4 nm

Once dye molecules have entered the intermolecular spaces, they are also held by forces of attraction (Gohl & Vilensky). Ion pairs or salt bridges are generally important for dyeing protein fibers (Christie, Gohl & Vilensky, Pyott, Trotman). In this process a charged dye forms a salt bridge with an oppositely charged group on the fiber. Excluding the positively charged anthocyanins, most natural dye chemicals do not have a significant charge at neutral pH and could not form salt bridges, however sulfate derivatives of flavonoids are quite common in leaves (Harborne 1979). The sulfates substituted on the hydroxyls imbue a negative charge at neutral pH to an otherwise uncharged dye resulting in a natural dyestuff capable of forming salt bridges. At acid pH commonly used in dyeing wool, the wool fiber has a net positive charge (Christie). The negative sulfate group of the natural dyes and the positive charge on the wool surface probably accounts for the affinity of a number of natural dyes for protein fibers.

Interestingly, many synthetic dyes have sulfate substituents to increase solubility and allow for ionic interactions with fibers. A salt bridge formed by interaction of a positively charged group on a protein fibers and a negatively charged dye is shown in fig. 7.5. Many natural dyes also have phenolic hydroxides which will attain a negative charge under alkaline conditions (i.e. pH > 9, Swain 1976) and could then interact by salt bridges. Strongly alkaline conditions can damage wool by hydrolyzing the protein to its component amino acids, however milder alkaline conditions can be tolerated to some extent (Aspland, Christie, Storey, Trotman). Increasing the pH of the dye vat will change the charge profile of the wool, making it's overall charge less positive.

Fig. 7.5 The ionic interaction between a negatively charged dye and a positively charged group in a protein fiber. These interactions are very important for acid dyes on wool and other protein fibers. These attractions are called salt bridges or ion pairs.

The formation of hydrogen bonds between fiber and dye chemical is important for natural fibers as well (Christie, Gohl & Vilensky, Pyott, Trotman). A hydrogen bond is not a bond, but a very strong attractive force, second only to salt bridges. Hydrogen bonds form by interaction of the lone pair electrons from the very electronegative atoms of O, N or F with an H bonded to another O, N or F. The molecule whose hydrogen is involved in the bond is called the hydrogen bond donor and the group whose lone pair electrons are involved is called the hydrogen bond acceptor. A simple hydrogen bond between two water molecules is shown in fig. 7.6. Hydrogen bonds are in reality an unusually strong type of dipole-dipole force discussed below. Because they are so much stronger than most other dipole-dipole interactions, they are usually treated separately.

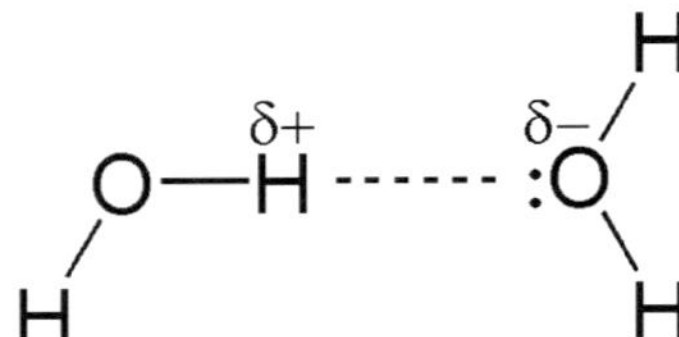

Fig. 7.6 A hydrogen bond between two water molecules. The water molecule on the left is the hydrogen bond donor because its H is involved in the bond. The water molecule on the right is the hydrogen bond acceptor because its lone pair electrons "accept" the hydrogen.

Natural dye chemicals have many hydroxyl groups capable of serving as both hydrogen bond donors and acceptors. Cellulosic fibers also have many polar hydroxyl groups available and a small number of these are ionized to negatively charged oxygens at normal pH's (Christie). Some carboxyl groups are formed by the oxidation of hydroxyls on carbon 6 of glucose (McGregor, Motomura *et. Al.*). Bleaching and other processing of cotton increases the number of carboxyl groups on the fiber (Motomura *et. al.*). As a result, cotton and cellulosics attain a slight negative charge on the surface of the fiber (Christie). Protein fibers have less hydroxyl groups but have amines, carboxyls and amides capable of participating in hydrogen bonds. These appear to be particularly important with direct or substantive dyes that are fast without a mordant (Gohl & Vilensky, Pyott). Hydrogen bonding is very important for the interaction of direct dyes with cellulosic fibers (Christie).

The category of interactions known as van der Waal's forces is important in the binding of dyes (Aspland, Christie, Gohl & Vilensky, Pyott, Trotman). Van der Waals forces are strongest with large molecules, in particular large planar molecules and the fastness of natural dyes appears to be related to their size to some degree. Van der Waal's forces include dipole-dipole attractions, induced dipole-induced dipole attractions and London or dispersive forces.

As already mentioned, hydrogen bonds are a special subset of dipole-dipole interactions. In general, dipole-dipole forces are electrostatic forces of attraction between *neutral* polar molecules. Neutral polar molecules or groups are best described as molecules that have no net charge but exhibit some charge separation within the molecule. Thiols or sulfhydryls are a good example of a polar group. The sulfur atom is partially negative while the hydrogen atom is partially positive resulting in a polar bond. Similarly carbon-halogen bonds are polar with the halogen partially negative and the carbon partially positive. These polar groups will interact with other charged or polar groups as shown in fig. 7.7.

Fig. 7.7 Polar molecules will be attracted by dipole-dipole interactions. a) a generalized dipole-dipole interaction. b) the dipole-dipole interaction between a sulfhydryl group and a carbon-chlorine bond.

Protein fibers contain polar amino acid residues including thiols. Most natural dyes have polar groups, often hydroxyls and cellulosic fibers have many very polar hydroxyl groups. Because of the abundance of groups capable of dipole-dipole interactions, they probably play some role in fixing natural dyes to fibers. They are undoubtedly important in attracting the dye to the surface of the fiber during the first stage of dyeing discussed above.

Induced dipole-dipole interactions occur when one charged or polar group induces charge separation in a neighboring group thus forming an induced dipole. The polar group and the induced dipole will enter into a dipole-dipole bond. London or dispersive forces are similar to induced dipole formation. In this case a neutral molecule will eventually attain a slight charge imbalance as a result of the movement of its electrons. This temporary dipole will induce polarization in an adjacent molecule. The original "temporary dipole" will not retain this polar nature and will quickly revert to a nonpolar arrangement of electrons, but by then it has already induced a dipole in its neighbor. Similarly the neighbor will induce a dipole in another neighbor and so on.

Van der Waal's forces are quite weak in general and depend on the surface area of the molecule. Larger planar molecules will have more surface available to interact with fibers, and planar vat dyes are more substantive than nonplanar dyes (Christie, Trotman). In the case of van der Waal's forces, a very large number of weak forces results in a strong interaction. The best analogy for this effect is hook and loop tape fasteners or Velcro™. Small fasterners are quite easy to open, the interactions between the hook and loop tapes being fairly weak. On the other hand, most students will have seen television or pictures of men wearing "Velcro™ suits" hanging from "Velcro™ walls." This analogy can reinforce the idea that a large number of weak forces can contribute to a significant net force.

The last important interaction of natural dyes with fibers is the hydrophobic effect discussed previously as a driving force for dyeing (Aspland, Trotman). The hydrophobic effect is an entropic effect caused by the removal of hydrophobic groups from water. Once a large, mostly hydrocarbon, dye molecule has migrated into the fiber the entropy of the system has been raised. The return of the dye to the bulk solution would mean a decrease in the entropy of the system, which is thermodynamically unfavorable and so equilibrium considerations favor the dye remaining in the fiber. Although the hydrophobic effect is a thermodynamic driving force for dyeing, it is also important in fixing the dyes to the fiber.

The Chemistry of Mordants and Mordant Dyes

Most natural dyes require a mordant to make the dye fast, or to fix the dye to the fiber. Many are weakly colored or nearly colorless without a metal ion but become highly colored when bound to a mordant (Swain, 1976). The most important mordants are all positively charged metal ions. Tannins are often used as mordants with cotton, but the chemistry is different and will be discussed separately from the metal ion mordants. Mordants are usually reported to act by forming a bridge between fiber and dye (Cannon & Cannon, Liles, Storey). Wool and other protein fibers will bind metal ions and retain them in the fibers (Androsko) presumably by ionic interactions between negatively charged amino acid residues of the protein and the positively charged metal ions. Once a dye is added to a mordanted fabric, an insoluble form, called a "lake," is produced (Christie, Gohl & Vilensky, Pyott, Storey, Trotman) trapping the dye in the fiber.

Studies with synthetic and natural mordant dyes have shown the nature of this insoluble lake. The dyes act as ligands and form complexes with the metal ions through the process of coordinate covalent bonding (Trotman). A coordinate covalent bond is formed when one atom contributes both electrons needed for bond formation as shown in fig 7.8. The resulting complex is either insoluble or at least significantly less soluble than the uncomplexed dye. If this process occurs within a fiber, the low solubility of the resulting complex makes it less likely to diffuse back out. In addition its large size will physically hinder migration out of the fiber (Christie, Gohl & Vilensky).

$$O{:} \quad + \quad M^{3+} \quad \longrightarrow \quad O{:}M^{3+}$$

Fig. 7.8 A coordinate covalent bond is formed when both electrons of the bond are donated by one atom. The most common electron donors are oxygen and nitrogen.

In one well-studied case the quinoid dye chemical alizarin was shown to form a complex with Cr(III). The complex is formed between three alizarin molecules and one ion of Cr (III). The alizarin ligands in this complex are bidentate, that is each alizarin is bonded to the Cr(III) in two distinct places. The carbonyl oxygen is bonded to Cr(III) through a coordinate covalent bond while a second hydroxyl oxygen is bonded to Cr(III) ionically via a salt bridge as shown in fig. 7.9 (Trotman). Alizarin has also been shown to form complexes with Al (Christie). Other mordant dyes have also been shown to form similar complexes with dye molecules (Trotman). It is expected that the metal ions complex with groups in the fibers and form a complex of dye, metal and fiber (Gohl & Vilensky, Pyott). Metal ions do bind to wool fibers. Cr (VI) oxidizes the amino acids cys, met and tyr and simultaneously is reduced to Cr (III) which binds to carboxyl groups of wool (Christie, Maclaren & Mulligan). Cu (I), Cu (II), Fe (III) and Al salts all bind to wool (Maclaren & Mulligan).

Although few studies have been conducted with natural dyes or with mordants other than chrome, it appears likely that a similar mechanism explains the attachment of the natural mordant dyes to protein fibers. Indeed although alizarin is now produced synthetically it is the dye chemical found in the natural dye madder and flavonoids are known to complex with metal ions in plants (Harborne 1979). Many natural dyes are colorless flavonoids that develop different colors by virtue of their interaction with different metal ions (Hendrickson). Hydrogen bonding and van der Waals forces are also important in dye color (Gohl & Vilensky, Pyott).

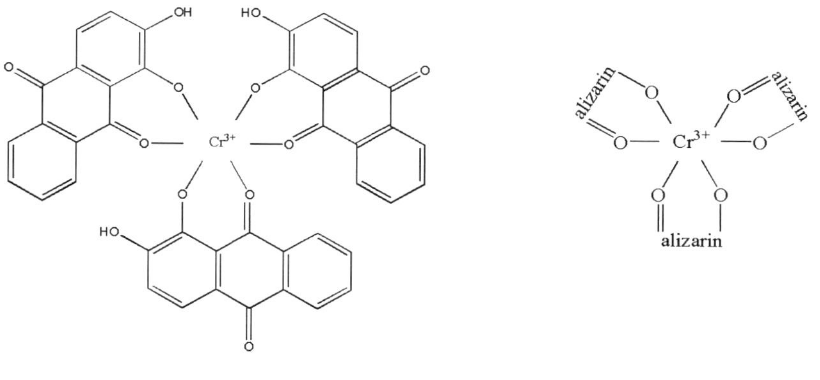

Fig. 7.9 The formation of a complex between alizarin and Cr(III). The complex is formed from one Cr(III) and three alizarin molecules. Each alizarin is a bidentate ligand, with one ionic and one coordinate covalent bond.

A similar mechanism undoubtedly explains the effect of metal ion mordants in the dyeing of cellulosic fibers. Cellulosic fibers, however, do not retain metal ions as wool does (Androsko) and thus require the addition of tannin mordant. Tannins are a class of large polyhydroxy compounds also referred to as polyphenols. Tannins are grouped into two broad categories based on structure and molecular weight. The condensed tannins are very large and do not contain sugar residues in their structures (Goodwin & Mercer). The smaller hydrolysable tannins have sugar cores, often glucose (Waterman, Goodwin & Mercer). The name "tannic acid" is often used as a synonym for hydrolysable tannins (King). These tannins are generally used as a mordant rather than the larger condensed tannins. One of the simplest hydrolysable tannins is shown in fig 6.15.

Cotton and other cellulosic fibers will interact with tannins, presumably through hydrogen bonding and van der Waals interactions. Both of these factors are known to be important in dyeing cellulosics (Christie, Gohl & Vilensky). Tannins can also associate with proteins by hydrogen bonding and the ability to strongly bind to proteins is characteristic of tannins (Waterman) The tannin protein hydrogen bonds are difficult to break at normal pH (Swain 1979). Tannins preferentially bind proline rich proteins (Waterman) which accounts for their ability to "tan" leather. Leather is primarily composed of the proline rich skin protein collagen. Tannins do not bind well to small molecular weight or water soluble globular proteins (Waterman) but will bind to other fibrous proteins such as those in wool and other protein fibers. The hydrogen bonds between tannin and protein can autooxidize to covalent bonds (Swain 1979). It is possible that a similar mechanism occurs with cellulosic fibers.

The tannins will bind metal mordants, possibly by complex formation. They will form colored complexes with iron salts (Haslem) as in the iron-tannate dyes (Chap 3). When used as a mordant, the tannins can serve as a bridge between the metal ion mordant and the cellulose. The metal ion mordant is now fixed in the cellulosic fiber and free to bind to dye molecules in the same manner as with mordanted protein fibers. The large size of the tannin dye complex would hinder migration out of the fiber.

In addition the tannins themselves may bind dyes without metal being required. Tea, onion skin and black walnut all contain large amounts of tannin and can function as direct dyes requiring no added mordant and flavonoids are commonly found associated with tannins *in vivo.* The flavonol fisetin is present as a tannin glycosides in sumac heartwood and the flavone morin is found as a tannic acid glycoside (Mayer & Cook). Quinoid pigments have also been shown to react with tannins (Chenyier) and this is certainly important in dyeing with naphthaquinones. Hydrolyzable tannins covalently bonded to anthocyanins have been found in wine (Cheynier).

Two mechanisms have been proposed for the formation of these covalent bonds between anthocyanins and tannins. The first is addition of the anthocyanin electrophile to nucleophilic catechols on condensed tannins (Chenyier). The second proposed mechanism is an acetaldehyde mediated condensation reaction requiring metal ion catalysis (Cheynier). It seems reasonable that similar mechanisms would explain the binding of anthocyanin dyes to tannin mordants, particularly in the presence of metal ions.

Ionic interactions are generally not considered important for cellulosic fibers (Christie). For the most part, this is based on the inability of negatively charged direct dyes to form salt bridges with negatively charged cellulosic fibers as opposed to those formed with positively charged wool fibers. When discussing natural mordant dyes however, the possibility of ionic interactions between fiber and mordant must be considered as well. It seems likely that the chelation of positively charged mordants accounts for the fact that bleached cotton attains a darker shade than unbleached cotton. The bleaching processes damage cellulose and as a result more of the primary alcohol groups (CH2OH) in cellulose are oxidized to negatively charged carboxylate groups (COO-). In fact a similar mechanism has been demonstrated with wool. Oxidatively bleached wool sorbs more aluminum ions than unbleached wool (Maclaren & Mulligan). The negatively charged carboxyls on the bleached cotton probably form salt bridges with the positively charged metal ions as they do in wool mordanting.

The hydroxyl groups of cellulose also ionize as a function of pH and ionic strength to become negatively charged (Motomura *et.al.*). While the pKa of 14 is rather high, the formation of ionized hydroxyls is necessary for their reaction with reactive dyes (Christie, Motomura *et.al.*). The ionized hydroxyl groups play an important role even though their pKa is high, primarily because of the large number of hydroxyls on cellulose and because ionization is an equilibrium process. As free ionized groups undergo reaction the equilibrium will shift according to LeChatlier's Principle and additional groups should ionize. This equilibrium accounts for the negative charge on cellulosic fibers in the dyeing process and may be important in binding mordants as well.

Dye binding and the actions of mordants is complex. The interactions are affected by a large number of factors. Most studies are conducted using commercially important synthetic dyes rather than natural dyes. Mordant dyes are usually studied only with Cr (III) mordant. As a result, interpretation of the results relevance to natural dyes can be difficult. Ongoing studies will continue to yield better understanding of this fascinating field and shed more light on the interesting chemistry of natural dyes.

Chapter 8
The Biology and Biochemistry of Natural Dyes

Natural Dyes come from animal, vegetable and mineral sources. The dyes presented in this book are primarily vegetable in origin as are the majority of natural dyes. Iron buff, a mineral dye, is an exception as are the red dyes derived from insect species including cochineal, lac insect, kermes, eucalypt scale insect and the purple dye derived from some species of shellfish. The mordants used in natural dyeing are primarily mineral substances. Although modern dyers purchase the metal salts ancient dyers used the raw minerals around them. This is clearly illustrated by the traditional Iron Buff, Iron Tannate and Goldenrod dye recipes provided in Chap. 3.

Novice dyers find out early in their dyeing careers that what you see in the plant is often *not* what you get in the pot. Many brightly colored flowers do not produce that same color in the dyepot and sometimes all color disappears completely. Other times weakly colored substances produce stunning colors. Not all colors are equally common either. Yellows, golds, tans and buffs are quite easy to attain. Orange, pink and pale red dyes are less common, but can be produced from some plants. On the other hand, plants that produce blue, purple and deep red dyes are rare even though these colors are quite common in flowers. Even more confusing in a world of plants with green foliage, is why a true green natural dye is essentially nonexistent (Trotman).

Inevitably this leads to several questions. Why are blue, purple and saturated red so common in flowers but so hard to get in the dyepot? Why are yellowish shades so easy to get even from things that aren't yellow in nature? Why do colors sometimes appear or disappear in the dyepot? Why do nut hulls, heartwood, leaves and roots have colored compounds in the first place? Why are the same dye chemicals found so commonly and in such taxonomically diverse plants while others are quite rare and restricted to one family, one genus or even one species? And why can't anybody find a green natural dye? Most or all of these questions can be answered by an understanding of the chemicals that make up natural dyes, the biochemistry of the chemicals and their distribution and functions in plants. With the background of Ch. 6 and 7 we can now answer these questions.

Anthocyanins and Anthocyanidins: Why are Blues, Violets and Deep Reds Hard to Get?

Nearly all red, blue and violet flowers and fruits are colored by anthocyanins (Harborne 1967). The red, blue and violet colors are so common because anthocyanins function in the flowers as attractors for pollinators and in fruits to attract animals who eat the fruit and distribute the seeds (Harborne 1976). They are found exclusively in the angiosperms (Harborne 1967), the most highly evolved plants. They probably co-evolved with pollinators and anthocyanin containing plants have colors consistent with pollinators' preferences (Harborne 1976 and1994, Swain 1980). Hummingbirds prefer reds and oranges like those of pelargondin (lambda max 520 nm), bees prefer blues like those of delphinidin (lambda max 546 nm) and butterflies prefer magenta and pink, like those of cyanidin and delphinidin (lambda max 535 nm and 546 nm respectively).

Anthocyanidins, the core components of anthocyanins, are colorless under neutral conditions, but undergo color changes as a result of the interesting acid-base chemistry shown in Fig. 8.1. Anthocyanidins are red or orange in acidic solutions, colorless in neutral solutions, and blue in alkaline solutions (Goodwin & Mercer, Mayer & Cook).

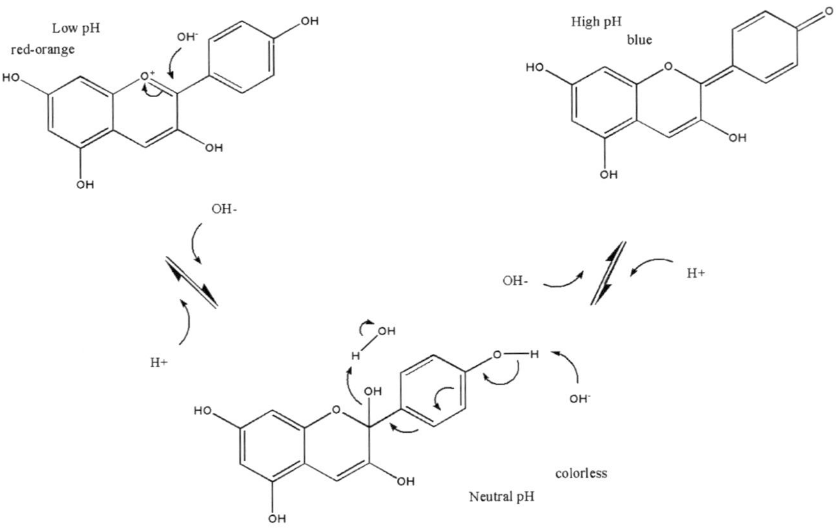

Low pH
red-orange
High pH
blue
Neutral pH
colorless

Figure 8.1 The acid-base chemistry of anthocyanidins results in reds and oranges at low pH, the absence of color at neutral pH and blue at high pH. The high pH form is called the anhydrobase. As interesting as this chemistry is, it is probably not a major factor in flower color. Adapted from Goodwin & Mercer and Mayer & Cook.

Originally everyone assumed that the pH effect explained the various colors of flowers. Indeed, cell pH can be a factor in determining the color of some flower families with Begonia's having a sap pH as low as 2.5 and blue morning glories as high as 7.5 (Hrazdina), however most plants have sap pH's at which anthocyanidins are colorless (Harborne 1976, Hrazdina) and other factors determine the color of the flowers. Even if it is rarely the most important factor in flower color, the pH dependence of anthocyanidin color is important for dyers to understand. The pKa of cyanin is about 2 (Swain1976) which means that cyanin will be reddest at pH below 2. Although other anthocyanidins may have somewhat higher pKa's (Swain 1976), some perhaps as high as 4, the production of a saturated red color requires acidic conditions. Adding vinegar to the dyebath or rinsing the dyed item in an acidic afterbath is often recommended (Buchanan, Hess) but this will not reduce the pH sufficiently to result in all the dye molecules being red. Acid base changes are equilibria, so when the pH equals the pKa half of the molecules will be in the red form and half in the colorless form. The higher the pH, the more molecules there will be in the colorless state. The resultant color will be pink or buffish pink. Many dyers are disappointed when published recipes for anthocyanidin dyes fail to produce vivid reds because they do not understand this pH dependence.

As mentioned above other factors are usually more important in flower color than cell pH (Goodwin & Mercer, Harborne 1976 and 1979, Hrazdina, Trevor Robinson). One of the simplest factors in red flowers is the attachment of sugars found on nearly all anthocyanidins *in vivo.* These anthocyanidin-sugar compounds or glycones are referred to as *anthocyanins* and the aglycone is an *anthocyanidin* (Swain 1976). The sugar on the anthocyanins helps keep them colored at sap pH. The type of sugar and the site of attachment vary from family to family (Harborne 1979). Interestingly, the identity of the attached sugar has little or no effect on the color of the anthocyanin but the position of the attached sugar can result in different colors for the same anthocyanidin core (Harborne 1976). In the extraction process, the sugars may or may not be hydrolyzed from the anthocyanins leaving behind the anthocyanidin core, which is only colored at low pH. Some red flowers and berry epidermal tissues actually do contain free anthocyanins (Harborne 1976, Hrazdina). In these red flowers and fruits, cell pH is probably an important factor in their color.

The most important factors in blue flowers are neither pH nor sugar attachment, but the chelation of metal ions by anthocyanins and/or association of anthocyanins with co-pigments (Goodwin & Mercer, Harborne 1976 and 1979, Hrazdina). Metal ion chelation by anthocyanins is responsible for the color in most blue fruits (Goodwin & Mercer). Aluminum and iron are the metals most commonly found in complexes with anthocyanins (Harborne 1976). Blue hydrangeas have an anthocyanin complex with iron and molybdenum (Harborne 1976). *In vivo* this metal chelation is dependent on the concentrations of a number of other metal ions in a complex manner (Harborne 1976). The metal ion complexes dissociate when extracted, particularly in acid pH's (Harborne 1976). So while the flower may be blue, the extract will most likely be colorless, thus blues are hard to get.

This effect seems less pronounced with berries. The metal complex may dissociate when extracted but solutions do retain some color indicating some interaction between the pigment and metal ions. The alum mordanted fabrics would be expected to bind these dyes to give blue to some extent, particularly because aluminum is one of the most important metals for producing blue *in vivo.* Natural dyers are often attracted to berries because of their ability to produce lasting stains on items of clothing.

Co-pigmentation of anthocyanins in flowers with flavonoid glycones and/or hydrolyzable tannins can also result in a blue colors (Goodwin & Mercer, Harborne 1976 and 1979, Hrazdina, Trevor Robinson). The co-pigments associate with the anthocyanin by hydrogen bonds between the carbonyl of the anthocyanin anhydrobase and aromatic hydroxyls of the flavonoids or tannin (Goodwin & Mercer, Harborne 1976, Hrazdina). There are also additional electrostatic and steric interactions between the anthocyanin and its co-pigment (Hrazdina). As can be seen in Fig. 8.1, the anhydrobase form of anthocyanins is blue so co-pigmentation with the anhydrobase stabilizes the blue form at pH's where it would not normally exist. The hydrogen bonds in co-pigmentation are pH dependent and fairly weak (Hrazdina). Co-pigments may be lost when extracting dyes. At the very least, the interactions are weakened with heat and the blue color of a dye often reddens as heated but returns to blue when cooling (Harborne 1979).

Some of the co-pigment arrangements can be quite complex. In blue cornflower the complex includes one iron ion, three apigenin derivatives (flavones) and four cyanidin molecules (anthocyanins) (Goodwin & Mercer, Harborne 1976). Commelina's blue color is caused by a complex of one magnesium ion, two molecules of an apigenin derivative (flavone) and two molecules of a delphinidin derivative (anthocyanidin) (Harborne 1967). These huge complexes would certainly be disrupted during the extraction process with the resultant loss of the blue color between flower petal and dyepot. Interestingly, the deep purple-black color of eggplant skin caused by delphinidin (Harborne 76) is not a result of metal ion chelation or co-pigmentation, but a stabilized anhydrobase (Hrazdina). It would be interesting to see if the eggplant color could be extracted into a nonaqueous solvent and used as a dye!

Carotenoids and Flavonoids: Why are yellows so easy to get?

There are many different pigments that can color flowers, fruits and roots yellow including carotenoids, flavonoids, anthocyanins, anthochlors and alkaloids. All of these can be important in natural dyeing. Carotenoids are the major pigment in most yellow flowers, particularly bright yellow flowers like buttercups (Harborne 1976). They can occur in petals in complex mixtures with as many as 15-20 different compounds in the same petal (TW Goodwin 1976). Mixtures of carotenoid and flavonoid or carotenoid and anthocyanin are responsible for the yellow and orange color of many flowers as well (Harborne 1976).

Carotenoids also color some yellow, orange and red fruits and vegetables (TW Goodwin 1976, Harborne 1976). Yellow cherries are colored by carotenoids (Harborne 1976) as are red and yellow peppers, citrus fruits, pineapples, papaya, watermelon (TW Goodwin 1966 and 1976). The red-orange color of ripe tomatoes is caused mainly by beta-carotene and lycopene and the familiar orange carrot is the most well known vegetable colored by the carotenoid beta-carotene. Sweet potatoes and corn seeds are also colored by carotenoids (TW Goodwin 1976).

Carotenoid pigments beta-carotene and lutein are always present in higher plants and indicate a common ancestry (TW Goodwin 1966). One reason for their occurrence in lower plants is their role in photosynthesis in these plants, however they serve a number of different functions in higher plants (TW Goodwin 1966). They are believed to protect plants against photodegradation of chlorophyll and to attract pollinators (Burnett). Their coloration of fruits is probably to attract animals that disperse the seeds (Harborne 1976). Old world monkeys are known to be attracted to yellow and orange colors typical for tropical fruits (Hutchings). They may also play a role in attracting pollinators (Burnett, Harborne 1976).

Many of the yellow dyes in this book are flavonoid dyes and flavonoid dyes are quite common. Interestingly, the yellows of fruits and most yellow flower petals themselves are *not* caused by flavonoids even though flavonoids are present in the flowers (Goodwin & Mercer, Harborne 1976). Flavonoids are nearly universal in higher plants, with fully 75%

of angiosperms having the flavonols quercetin, kaempferol and/or myricetin (Harborne 1967). Quercetin and kaempferol are very primitive pigments (Swain 1980) and their presence in so many diverse species is a result of common ancestry (Heywood). Even most white and cream flowers have flavonol or flavones (Harborne 1976) that appear nearly colorless as glycosides (Trevor Robinson).

Free flavonoids can be found in non-living heartwood of trees (Swain 1976). Although flavones and flavonols both have their maximum absorbance in the ultraviolet region the absorbance bands are broad. As a result at high concentrations there is sufficient absorbance in the blue region to produce a yellow color (Swain 1976). In most plants, flavonoids occur as glycosides (Harborne 1994, Swain 1976). The glycosides are more soluble and more chemically stable than the free flavonoids (Swain 1976) and also less colored than the parent flavones and flavonoids (Trevor Robinson). Most plants have more than one glycoside of the same pigment. For example huckleberry has five different quercetin glycosides (Swain 1976). Quercetin is the most widely distributed of all yellow pigments (Robinson). At least 70 and perhaps as many as 100 different glycosides of quercetin have been identified in plants (Harborne 1979, Swain, 1976).

Some flavonols do contribute to flower color, particularly quercetagetin, the orange yellow pigment responsible for the color of marigold species, and methylated flavonol derivatives (Cannon & Cannon, Goodwin & Mercer). The yellow flowers of primrose and cotton plants are also colored by flavonols (Harborne 1976). In most cases, however, the flavonoids are believed to serve as nectar guides for bees (Harborne 1979, Swain 1976). While the flavonoid glycosides do not absorb appreciably in the visible light range, they absorb strongly in the ultraviolet region (Harborne 1976). The wavelengths of maximum absorbance for the various flavonols and flavones, lambda max, are in the range of 330-375 nm (Trevor Robinson). Honey bees have a vision receptor with a lambda max of 340 nm (Hutchings) which corresponds to flavonoid absorbance band. Bees see the absence of reflected ultraviolet "light" as dark areas directing them to the nectar.

Other important functions of flavones and flavonols in plants may be protective. Flavonoids may serve as both insect attractors and insect repellents (Harborne 1979, Swain 1979). The famous preferrence of the silk worm for mulberry leaves is a classic example. The silkworm is attracted by the flavone morin found in the mulberry leaves while many other insects are repelled by morin. If the position of the sugar attachment to morin is changed, the silkworm will not feed either (Harborne 1979). It has been suggested that flavonoids may protect against phytopathogens, deter herbivores from grazing (Swain 1980) and serve as ultraviolet lightscreens (Swain 1979) but the evidence is primarily circumstantial (Hrazdina).

In view of the ubiquitous distribution of flavones and flavonols in plants, it is not surprising to find a yellow dye in the dyepot even if the flower or wood was not itself yellow. In the absence of other colored pigments, the aglycone yellow color will appear. This is particularly true in alkaline solutions where even the glycosides will appear more brightly yellow and orange (Trevor Robinson). Blue flowers in particular often result in pale yellow dyes. Blue flowers are usually a result of co-pigmentation of anthocyanins and flavonoids as described above. During extraction the co-pigmentation is usually disrupted and the resulting free anthocyanin reverts to the colorless form. The free flavonoid co-pigment yields a pale yellow solution and if you are lucky, a good yellow dye.

The yellow pigments with perhaps the most fascinating chemistry are the anthochlors. Like the anthocyanins, these pigments change color with pH. The anthochlors include chalcone and aurone pigments. Chalcones and aurones are often referred to as flavonoids. While strictly speaking they are not flavonoids, they are closely related (Goodwin & Mercer, Swain 1976) and are usually included with the discussion of flavonoids (Swain 1976). Unlike the flavonols and flavones, anthochlors are less widely distributed, occurring mostly in composite flower petals although chalcones can be found in heartwood, fruits, leaves and bark (Harborne 1967). Aurones are more restricted in their distribution, generally found in a few families of composites (Boehm).

The pigments were dubbed anthochlors because their normal yellow colored petals become quite red when exposed to ammonia vapor (Trevor Robinson, Swain 1976). The absorbance spectra of all flavonoids are markedly changed at higher pH as a result of ionization of phenolic hydroxide residues (Swain 1976) and a shift of the absorbance maximum toward longer wavelengths. The yellow to red color change is only observed with the anthochlors, however. Chalcones have lambda max between 370-390 nm and aurones between 390-420 nm (Harborne 1976) while flavones and flavonols have lambda max between 330-375 nm (Trevor Robinson). The maximum absorbance of anthochlors in or very near the blue end of the visible range, rather than the ultraviolet region, means that a red shift moves the absorbance into the visible region. This results in a color change rather than just a deepening or weakening of the yellow color as is seen if the red shifted absorbance remains in the ultraviolet region. It also explains the brighter yellow color of the anthochlors pigments relative to flavonoids. In spite of their bright yellow color, like flavonoids, anthochlors appear to function as nectar guides, appearing dark to honey bees relative to the carotenoid pigments they are found in conjunction with (Boehm).

Why do colors disappear, change or "magically" appear in the dyepot or the fiber?

The "disappearing color" is a particular problem for blue flowers which are nearly always colored by anthocyanins. As noted above, the most important factors in blue anthocyanins are the chelation of metal ions and the association with co-pigments (Goodwin & Mercer, Harborne 1976 and 1979, Hrazdina). The metal ion complexes responsible for the blue color in flowers such as Hydrangea dissociate when extracted, particularly in acid pH (Harborne 1976). At the near neutral pH of the dyebath, the free anthocyanin will revert to the colorless form. Anthocyanins are destroyed by air at alkaline pH (Trevor Robinson) so making the dyebath alkaline will not help preserve blue color either. This results in the loss of the blue color between flower petal and dyepot. Similarly, co-pigments may be lost when extracting dyes. As described above, the associations between anthocyanins and co-pigments can be very complex and the interactions fairly weak. Even if the blue color is maintained in solution as with blueberries, it may be too big to penetrate the fibers effectively. If the dye cannot migrate into fiber, it will not be a fast dye.

"Changing colors" are often a pH effect. Anthocyanin red dyes will be redder in acidic solutions and colorless in neutral solutions. Similarly the anthochlors are affected by pH and will change from yellows to reds in alkaline solutions. The chemistry of both processes is described above.

Other times colors that were not evident in the plant or flower "magically appear" in the dyepot during extraction. This is often true with flavonoid yellow dyes as described above and it can also be true of quinoid dyes. Quinoids are widely distributed in the plant kingdom and are produced by multiple biochemical pathways (Leistner). There are over 500 quinones found in plants, mostly in the wood, bark and root (Thomson 1976 and 1978). Quinoids are usually present as colorless glycosides (Goodwin & Mercer, Thomson 1978). Likewise, free glycosides that are in the chemically reduced state are also colorless (Goodwin & Mercer).

When extracted, reduced forms are quickly oxidized by air and the glycosides most likely hydrolyzed. Both these chemical changes result in the appearance of colors. A good example is the green hull of the black walnut. The hulls contain large amounts of colorless reduced pigment alpha-hydrojuglone (Goodwin & Mercer, Mayer & Cook) and when cut or pierced a pale yellow liquid is observed. Very quickly thereafter, a dark brown color appears as the reduced compound is quickly oxidized in air to the quinone, juglone. A similar but more dramatic process in the production of the anthraquinone dye China Green is described below.

While most quinoids range in color from yellow through red (Thomson 1976), some are reported to be green, blue or nearly black (Thomson 1978). In heartwood colored dimers, trimers and tetramers of oxidized quinines can be found (Goodwin & Mercer). The colored heartwoods of many trees are a result of these quinone oligomers.

A very complex example of air oxidation of pigments is the development of the blue color of indigo dyes. The pale yellow solution produces a greenish yellow fiber that slowly turns blue upon exposure to the air. This chemistry is explained in detail in Ch. 3. Similar chemistry is responsible for the colors of other vat dyes.

Why do nut hulls, wood, leaves and roots have colored compounds at all?

Nut hulls may be colored to attract animals who will carry the seeds away (Harborne 1976). Ruminants, squirrels and other rodents are attracted to dull green and brown fruits (Hutchings). Leaves may have these compounds to attract insects, as described for the silkworm above. Color may or may not play a role in this attraction.

Both flavonoids and quinoids may be present in wood and leaves in protective roles. Flavonoids may be lightscreens (Swain 1979), insecticides and herbivore deterrents (Harborne 1979, Swain 1979) as explained in detail above. Quinoid compounds probably also play protective roles in plants (Leistner). They may be antibacterial, antifungal and insecticidal (Leistner, Thomson 1978). Anthochlors are also protective. They appear to have some activity as larval growth inhibitors and to function as alleopathic agents, inhibiting growth of new seedlings too close to the parents (Boehm).

Why are some dye chemicals in so many plants, while others are quite rare?

In most cases the answer to this question is common ancestry and evolution. Phylagenetically diverse plants share compounds that are found in the oldest plants, namely flavones and flavonols. Flavonoids compounds increase in sophistication in higher plant species, but most plants have flavonols as discussed above. Generally flavonols are replaced by flavones and increased glycosylation is evident in more highly evolved plants. Flavonoids of increasing complexity are present and some are only found in angiosperms (Swain 1980). Other compounds, like the aurones are fairly rare (Swain 1976) and only found in composites (Boehm). These compounds occurred later in the evolutionary sequence and thus are only found in more closely related plants. Similar arguments can be made for other dye chemicals as well. Those that occurred earlier in the plant evolutionary process will be found in most plants. Those that developed later will be more restricted.

Why are there no green dyes?

Finally, the question about the elusive green dye, is probably the simplest question to answer. First, strictly speaking, it is not true. There are a few examples of green natural dyes however they are quite rare. Chap. 6 explained that dyes appear to be a color because they absorb light that is the compliment of the color. Table 6.1 showed that purple or magenta is the compliment to green (Griffith). Magenta is not a spectral color and thus green dyes are difficult to produce, even synthetically. To function as a green dye, a molecule must absorb both the red and blue ends of the spectrum (Griffiths, Trotman). The green pigment in plants, chlorophyll, does in fact absorb both red and blue light. Synthetic green dyes usually have a blue and a yellow dye molecule linked together chemically as described in Ch. 6.

Chlorophyll is certainly a naturally occurring green pigment but is not usually considered a green natural dye. This is no doubt due to its large size and low solubility in water. A method has been published for the extraction of chlorophyll into methylated spirits that are later diluted with water to produce a green dye (Smith). This method is not appropriate for children but may be of some interest for an organic chemistry class. The extraction of chlorophyll from spinach with organic solvents is a common organic chemistry laboratory experiment. While chlorophyll extracted and used in this manner may be considered a natural green dye, it does not have historical importance nor is it used by many modern hobbyists.

The Hawaiians are reported to produce a green natural dye using the berries *and* leaves of p polo, black nightshade plant *and* ma'o, native Hawaiian cotton (Krohn). This may be a confirmation of the difficulty in producing a green dye. The fact that the dye requires two plants including both leaves and berries of one argues for the presence of more than one dye chemical. Most berries are colored with anthocyanins (Harborne 1976) while the leaves probably yield tannins and yellow flavonoids and cotton flowers are known to be colored with flavonols (Harborne 1976). Over-dyeing with blue and yellow dyes was the prescribed way to produce green dyes with natural dyes in Europe and was still used for decades with early synthetic dyes (Brackman). So, the Hawaiians *do* produced a natural green dye but they do so most likely by extracting two or more dye chemicals into the same pot. This in an ingenious and definitely more efficient method than that used by European cultures, but it is probably not a single green dye chemical.

The only truly green natural dye may be Vert-de-Chine (Cannon & Cannon) or Chinese Green (Ponting) used in China in the 18th century. The dye is extracted from the bark of *Rhamnus utilis* (Cannon & Cannon, Ponting). Ponting lists Lo-Kau as another name for

China Green, but Cannon & Cannon report the dye called lokau to be a blue dye extracted from other rhamnus sp. According to Ponting, the China Green dyebath is colorless and the green develops upon exposure of the dyed fibers to the air. French chemists eventually produced a similar dye by boiling the bark of related species *R. catharticus* in the presence of large amounts of alkalai (Cannon & Cannon). The dye chemical in both cases is believed to be the anthraquinone dye rhamnicoside (Cannon & Cannon). As stated above, quinones can have a range of colors including green (Thompson 1978). Most anthraquinones dyes are vat dyes.

Chapter 9
The Color-fastness of Dyes

The methods of testing dyed fabrics for color-fastness used by industry are easy to understand and a good way to incorporate "real life" applications into the science curriculum. The modification of dye molecules by light is interesting chemistry as is the breakdown of fabric on light exposure. The degradation of dyes by washing depends on the chemistry and physics of the dye attachment to the fibers and the chemistry of the soap or detergent as well as the temperature of the wash. Modified versions of the same types of tests used to test color-fastness are easy to do in the classroom. Student experiments based on these tests can be found in Section 4.

No dyes are completely fast and the only reliable way to maintain color is to store the item in conditions of low humidity, low temperature and no light (Nassau c) which is not a practical solution. The fastness of synthetic dyes on cotton and wool can serve as a good comparison when talking about fastness of natural dyes. Wash fastness is judged on a scale of 1-5. Both wool and cotton dyed with reactive dyes have wash fastness of 5 while cotton dyed with synthetic vat dyes are slightly less fast with a rating of 4-5. Light fastness is judged on a scale of 1-8. Wool dyed with reactive dye has light fastness of 6, cotton dyed with reactive dye has 5-6, and cotton dyed with synthetic vat dye has 6-7 (Thompson and Thompson). In general, natural dyes will be less fast than synthetic dyes, but some are quite hardy.

Color-fastness depends on a number of factors and includes fastness to electromagnetic radiation (light-fastness) and fastness to laundering (wash-fastness) as well as the day to day conditions to which the item will be subjected, such as temperature, humidity (Csepregi *et. al.* 1998 b,Trotman) and pollution (Nassau c). Dyed items are tested for many of the conditions encountered by consumers who will purchase them (Storey) as well as those important for manufacturers (Trotman). These tests of importance to consumers may include fastness to sunlight, artificial light, exposure to the elements (Storey), pollutants like ozone (Nassau c) or conditions of extreme temperature and humidity (Trotman). These hazards might be encountered by any number of dyed items from clothing to umbrellas.

Similarly fastness to washing at various temperatures for various lengths of time and to dry cleaning (Pyott, Storey) must be tested for clothing items. Clothing must also be tested for fastness to pressing, rubbing, and perspiration and swim wear and beach items must be tested for fastness to seawater (Storey, Trotman). Items to be washed must be tested against bleeding onto other fabrics in the washer (Storey). The methods for conducting these tests can be somewhat complex. To ensure that a fastness rating will be reliable, stringent controls are used (Trotman). Some of these tests are summarized below along with the controls that are used to ensure reliable results.

Fastness to Light

The interaction of light and matter to produce a chemical reaction is known as photochemistry. The reactions that cause fading of dyes are photochemical reactions. There are two fundamental laws of photochemistry. First, the Grotthus Draper Law says that light must be absorbed to cause a chemical change. Second, the Stark Einstein Law says that a single photon of light must be absorbed (Cox & Kemp). So only the wavelengths of light that the dye can absorb need be considered as potential causes of photofading. This being said, however, all light is not equal.

The fastness of a dye also depends on both the type of light and length of time the fabric is exposed to that light. The amount of damage done by light exposure depends on the energy of the light. Ultraviolet light is the most damaging but violet and blue visible light also play a role in photo-degradation of dyes albeit more slowly (Nassau c). Windows offer significant protection against fading by sunlight and plastics such as Plexiglass are better. Sunlight through glass causes about 34% of the fading expected from direct sunlight and through plastic only 10% (Nassau c). Artificial light is less damaging than sunlight. Incandescent light causes about 3% of the fading expected from direct sunlight and fluorescent about 10%(Nassau c).

There are several different processes of photo-degradation of organic molecules (Cox & Kemp, Wayne & Wayne). Reactions include photoreduction, photoaddition and photoxidation (Wayne & Wayne). Photoreduction can occur when hydrogen atoms, H• radicals are abstracted from solvent (Cox & Kemp, Wayne & Wayne). The reduction of alkanes can occur by a mechanism of this type (Turro) as can the reduction of quinones (Cox & Kemp) and ketone carbonyls (Cox & Kemp, Turro). Synthetic azo dyes will undergo photoreduction of the azo bond under some circumstances, particularly in the presence of perspiration (Aranyosi 1998, Csepregi 1998b). Both cotton and wool fibers are themselves damaged by ultraviolet radiation resulting in hydrolytic reactions on the surface of the fibers (Gohl & Vilensky).

Photoreduction can also occur when light interacts with fiber rather than solvent to produce hydrogen radicals. This damages the fiber and the resulting hydrogen radicals attack the double bonds in the dye (Nassau c). Dyes are colored because they are conjugated (see Ch. 6). Reducing a double bond in the molecule destroys the conjugation and thus the color of the dye as shown in fig 9.1. This mechanism of fabric induced photoreduction is particularly important in silk items (Nassau c).

$$\text{H source} + \text{uv light} \longrightarrow \text{H}^{\bullet}$$

$$2\,\text{H}^{\bullet} + \text{Dye}\sim\sim\underset{\text{H}}{\text{C}}{=}\underset{\text{H}}{\text{C}}\sim\sim \longrightarrow \text{Dye}\sim\sim\underset{\text{H}_2}{\text{C}}{-}\underset{\text{H}_2}{\text{C}}\sim\sim$$

Fig. 9.1 The photoreduction of dyes can be caused by hydrogen abstraction from solvent or fabric. The reduction of the double bond will result in change in color or loss of dye color.

The second type of photochemical reaction mentioned above is the photoaddition of small electrophiles to alkenes (Wayne & Wayne) but there is no indication this mechanism is important with dyed fibers.

The third mechanism is oxidative photofading or photoxidation. In the ground state, molecular oxygen occurs as a triplet, meaning it has two unpaired electrons. In the photoxidation mechanism, triplet oxygen is photosensitized to the more reactive singlet oxygen (Cox & Kemp, Wayne & Wayne). The singlet oxygen now attacks other molecules. The photoxidation process is particularly important in unsaturated compounds (Cox & Kemp, Richard Wayne). Carotenoids in the chloroplasts of plants protect against sun damage to chlorophyll by reacting with and thus quenching the singlet oxygen (Wayne & Wayne). Photoxidation is also important in the light fading of azo dyes in the absence of reducing agents (Aranyosi *et. Al.* 1998, Csepregi *et al.* 1998 a, b) where dissolved oxygen acts as an oxidizing agent (Aranyosi *et. Al.*1999).

Nassau describes the photoxidation of dyed fabric in the presence of water. In this mechanism, water is attacked by oxygen and light to produce hydrogen peroxide (Nassau c). Indeed hydrogen peroxide radical is often the first product of photoxidation which then causes secondary reactions with the dyes (Wayne & Wayne). In Nassau's mechanism, shown in fig. 9.2, the peroxide adds to the double bond, oxidizing the carbons and converting the double bond to a single bond. Once again, destroying the conjugation of a dye will change or destroy the dyes color.

$$2\,H_2O + O_2 + \text{uv light} \longrightarrow 2\,H_2O_2$$

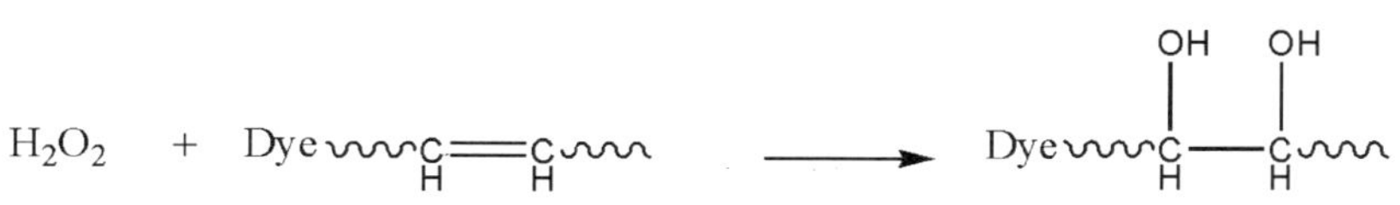

$$H_2O_2 + \text{Dye}\sim\sim\underset{\text{H}}{\text{C}}{=}\underset{\text{H}}{\text{C}}\sim\sim \longrightarrow \text{Dye}\sim\sim\underset{\text{H}}{\overset{\text{OH}}{\text{C}}}{-}\underset{\text{H}}{\overset{\text{OH}}{\text{C}}}\sim\sim$$

Fig. 9.2 The photooxidation of dyes caused by light and oxygen. The oxidation of the double bond will result in change of color or loss of dye color.

The mechanisms of dye photofading are often quite complex (Aranyosi *et. al.* 1998, Csepregi *et. al.* 1998 b). Photo-degradation is more typical of dyed fibers than of the dye molecules alone (Aranyosi *et. al.* 1998). Photo-fading of dyes can occur by photoreduction, photoxidation or even by mixed mechanisms (Aranyosi *et. al.* 1998, 1999, Csepregi *et. al.*

1998a, 1998b, Nassau c) Which process predominates depends on a number of factors including the nature of the fibers (Aranyosi *et. al.* 1998), and the type of dye-fiber bond (Aranyosi *et. al.* 1999, Csepregi *et. al.* 1998 b). Covalent bonds between dye and fiber increase the light stability. (Csepregi *et. al.* 1998 b). This undoubtedly explains why synthetic reactive dyes, that form covalent bonds to fibers, are more light-fast for cellulosics than natural dyes. The structure of the dye is also important in determining whether photoreduction or photoxidation will occur. In studies with synthetic azo dyes, the positions of substituents on the dye molecule and the nature of the chromophore itself, not the fiber, were the most important factor in determining the mechanisms of photo-fading (Csepregi *et. al.* 1998 a)

No matter what the fading mechanism, the number of molecules of dye affected by light is constant for a chromophore with identical exposure and is irrespective of the concentration of the dye in the garment (Trotman). As a result, pale and pastel items will be more affected, fading proportionately more than deeper hues (Trotman).

Testing for Fastness to Light

Light-fastness is tested on an ISO (International Organization for Standardization) scale of 1-8 with 1 corresponding to poor light-fastness and 8 corresponding to the maximum light-fastness (Storey, Trotman). Typical light-fastnesses for cotton and wool dyed with synthetic reactive dyes are 5-6 (Thompson & Thompson) corresponding to good to very good fastness and cotton with vat dyes 6-7 (Thompson & Thompson) corresponding to excellent fastness.

It is impossible to set up a uniform control for light exposure and so all tests must be run against a set of standard dyes of varying fastness and numbered sequentially from 1-8. The fabrics are set up with the sample and the standards all partially covered. The fabrics must be protected from rain by glass but air circulation must be maintained. The sample box is placed at an angle approximately equal to the latitude of the testing site and facing the direction of the sun, south in the northern hemisphere and north in the southern hemisphere (Trotman). As fading progresses a series of coverings are added to produce a series of increasingly faded bands. The light-fastness of the sample undergoing testing is rated by giving it the same number as the corresponding standard sample that exhibits a similar degree of fading (Storey,Trotman). If the series of tests yield different fastness ratings, the individual ratings are averaged and the mean is reported as the rating.

Sun fading is a slow process (Wingate) and most industrial applications use a xenon arc lamp to speed up the process (Thompson & Thompson). The lamp works more quickly and reliably than sunlight but care must be taken to filter long and short wavelengths to mimic sunlight accurately and prevent heating by infrared light that can cause heat induced fading (Trotman). A simplified version of the sun-fading system would be easy to build and use with students. Rather than authentic industry standards, the natural dyes could be compared to a series of synthetic dyes in purchased fabrics. A student experiment based on this test is included in section 4.

Fastness to Washing

There are many factors that affect the fastness of a dye to washing. These include properties of the dye such as its water solubility and chemical stability as well as conditions of the cleaning such as soap or detergent pH, the addition of bleaching agents to the detergent, the temperature of the wash, the type of water used and so forth (Nassau c). Two factors are important to test, whether or not the dye fades and whether or not the dye bleeds. Both of these can be tested simultaneously if the fabric to be tested is sandwiched between two other pieces of fabric of the same size and sewn together around the edges. The composition of the covering fabrics used with cotton or wool is one piece of cotton and one of wool. (Trotman).

There are several different methods used to test wash-fastness that increase in severity (Pyott, Trotman). The mildest of these uses water at 40°C / 104°F and 5g/L of soap for a 30 min. cycle, while the most rigorous uses 60°C / 140°F water and has sodium carbonate added to the detergent. The samples are rinsed, and the sewing is removed on three sides before drying at 60°C / 140°F (Trotman). The amount of fading is assessed by comparison with a 5 step grayscale (Storey, Trotman). One half of the grayscale (dark gray) is meant to represent the original fabric and the second half the faded fabric (Trotman). Bleeding into adjacent fibers is tested by comparing the sandwich fabrics to a different gray scale. On this grayscale the constant half of the scale is white, rather than dark gray (Trotman). Once again these methods would be easy to mimic with students and other factors, such as varying the detergent used and adding "color safe bleach" and other laundry treatments would be interesting extenstions.

Fastness to Other Factors

One of the most damaging chemicals for dyes is ozone (Nassau c). Ozone is an ingredient in photochemical smog experienced in cities. Ozone absorbs ultraviolet radiation and decomposes into chemically reactive products. These products react with the double bonds in dyes, destroying their color. Other pollutants can affect colors as well but are more important for pigments in painting and other art work than for dyes. Peroxides, particularly hydrogen peroxide will also destroy dye color although more slowly (Nassau c). Peroxides can be found in some laundry pre-soaks and additives, particularly those advertised for stain removal and disinfection.

Perspiration can affect dye fastness as well (Aranyosi *et. al.* 1998, Csepregi *et. al.* 1998b, Storey, Trotman). Perspiration or "sweat" is not just water. It contains salt and amino acids and varies in pH. Dyes are tested for perspiration fading commercially by using two "artificial sweat" formulations of different pH's. Items to be tested are soaked in one artificial sweat solution and placed between glass or plexiglass plates for 4 hours in an oven set at body temperature (Pyott, Trotman). The sample is dried and subjected to the same treatment with the second sweat mixture. A variation of this method is included as a student experiment in section 4. Sweat often contains lactic acid and other reducing agents and can increase the photo-fading of dyes as well (Aranyosi *et. al.* 1998, Csepregi *et. al.* 1998b).

Crocking or rubbing off of dye is also tested commercially by using a special machine that rubs the fabric rapidly and then is examined for crocking (Trotman). While this test would not be easy to mimic with students, most of them will have experienced it themselves. Vat dyes such as indigo are particularly subject to crocking because the dye molecules are attached at the surface of the fibers. Many other clothing items will show signs of crocking as well, often because of poor dyeing practice or failure to adequately remove unbound dye from the fiber (deBoer). Crocking is often visible as whitening at the seams of blue jeans and as coloration imparted to backpacks, purses or even other items of clothing that rub against the offending item of clothing.

Items of swim wear must be tested for fastness to sea water. This test is easy to mimic by repeatedly soaking the sample in saline solutions and then drying it. A similar test for chlorinated pool water might be more interesting to students not living in coastal areas. Most of them will have experienced swimsuits that faded over time, particularly if worn in highly chlorinated public swimming pools. Students can be encouraged to combine factors, such as sunlight exposure of the wetted samples, which would be common at beaches and pools, to make more complicated but more realistic extensions of the experiments in section 4.

Part 4
Natural Dyes in the Classroom

The classroom applications for teaching science with dyes are almost limitless. Incorporating examples of the chemistry of natural dyes into high school and college chemistry and biology lecture sections can add "real life" relevance for the students. Some possibilities include using the chemistry of indigo and other vat dyes to illustrate the concepts of oxidation and reduction, using the anthocyanin and anthochlors pigments when discussing acid/base chemistry of organic molecules, and using the anthochlors to illustrate the effect of ionization state on molecular absorbance spectra. These topics can be discussed simply as classroom examples or can be presented as laboratory experiments involving the dyes. There are many activities simple enough for primary school children and yet there are complex organic and physical chemistry experiments that can be conducted by college and university students. Some examples have been mentioned in the preceding text and more are presented in this section.

This section of the book presents a number of different activities and experiments appropriate for differing age levels and classes. Some ideas for integrated curricula and courses are suggested and extensions of some experiments are presented. Additional examples are included for older students as are ways of simplifying some activities for use with younger students. The final part of the chapters is a "bibliography" of additional dye experiments published elsewhere that may be of interest to teachers using natural dyes in the classroom.

Chapter 10
Activities, Experiments and Student Laboratory Sheets

Activities for Integrated Curricula

Teaching an integrated curriculum has become quite common in primary schools and including science activities is very important. However, it is not always easy for a teacher to find science activities suitable for an integrated curriculum, and few elementary teachers have the time or science background required to create their own activities. Natural dyes provide a number of opportunities to integrate science topics with history, social studies and art. Several activities suitable for integrated curriculum follow.

The Dye Day

A "dye day" is a great activity for an integrated history or social studies curriculum. It can be used with the youngest pupils up through middle school students. Classes studying pioneers, for example, can learn about the history, culture and science of the pioneers in an integrated style through a dye day activity. Cultural and scientific information about the dyes is given in this text. Teachers can easily add this information to the history and social studies curriculum. Depending on the age of the students, the number of classroom helpers, and the amount of time available, the dye day may be conducted outside in pots over campfires or inside the science lab using hot plates. Some dye palettes suitable for dye days highlighting various cultures are presented in table 10.1. Others themes are possible as well.

Table 10.1 Suggested dyes for use in Dye Days of an integrated curriculum. Other dye days are possible. Information about the cultures using the dyes can be found in the text. The chapters for information about the dyes are given in parenthesis.

Early American Settlers	Pioneer Days*	Native Americans#	African Peoples	Around the World
Simple iron buff (3)	Simple iron buff (3)		Simple iron buff (3)	
Slow iron liquor (3)	Passive iron buff (3)	Passive iron buff (3)		Passive iron buff (3)
	Iron Mud (3)	Iron Mud (3)	Iron Mud (3)	Iron Mud (3)
Sumac iron-tannate (3)		Sumac iron-tannate (3)		Sumac iron-tannate (3)
Generic bark dye (3)	Generic bark dye (3)	Generic bark dye (3)	Generic bark dye (3)	Generic bark dye (3)
Indigo (3)	Indigo (3)	Indigo (3)	Indigo (3)	Indigo (3)
	Onion skins (3)		Onion skins (3)	Onion skins (3)
Goldenrod yellow (4)	Goldenrod- Hopi (4)	Goldenrod yellow (4)		
	Anthocyanin dyes (5)	Anthocyanin dyes (5)	Anthocyanin dyes (5)	Anthocyanin dyes (5)
		Marigold dye (3)	Marigold dye (3)	
Black oak wood (4)	Black oak wood (4)	Barberry wood dye (3)	Barberry wood dye (3)	
Black walnut brown (4)	Black walnut brown (4)	Black walnut brown (4)		
	Black walnut, baskets (4)			
	Osage orange wood(4)	Coreopsis dye (4)	Weld (5)	
	Tea bag dye (5)	Sumac berry yellow (3)		

*The term "pioneers" here is used to refer to non-indigenous settlers in post-Colonial US. Included in this broad category are Appalachians, Emancipated African-Americans and various European settlers to the frontier regions. Although many of these dyes were used by most of these peoples, not all recipes are equally relevant to all groups. The text should be consulted for specific use of a dye by specific people.

#The term Native American here is used to include the indigenous peoples of the United States and North American including Alaska, Hawaii, continental USA, Canada and Mexico. Not all recipes are appropriate to all cultural groups within this broad heading. Details of use by various peoples can be found in the text.

Traditional Crafts and Household Arts

Two traditional crafts that involve natural dyes are Flower Pounding and Basket Making. In flower pounding, natural dyes are applied to mordanted fibers in selected patterns to create artistic designs. Flower pounding is described in detail and a method provided in Ch. 5. There are several books about flower pounding available. Other traditional crafts can be adapted to use natural dyes, such as needlework and quilting. Fabrics, yarn or embroidery floss can be dyed by the students and used for their projects. Dyeing methods and recipes are presented in Ch. 2-5.

Baskets are made all over the world in nearly every culture. Many peoples used natural dyes to color their baskets. The tradition of dyeing baskets with natural dyes includes the Appalachians with black walnut, Alaskan Tlingits with berries (probably anthocyanins) and roots (probably quinoids), Navajo Indians with various dyes, the Aboriginals of Queensland Australia with roots (probably quinoids) and the Kikuyu people of Kenya Africa with roots (probably quinoids). Details of these uses can be found in the text along with a recipe for using black walnut to dye baskets. Students can easily weave baskets from pliable materials such as raffia. Recipes written for cotton dyes can be used with basketry materials, particularly raffia and similar materials that can be simmered before weaving (Navajo school).

Household arts can include many uses of natural dyes. Tie dyed t-shirts using natural dyes are always popular with children. Antiquing of doilies and other household items is described in Ch 5. Natural dyes derived from food products can be used as food dyes provided only cooking pans and food products are used. Some examples are dying eggs with modified tea-bag dye recipes provided in Ch 5 by adding a little vinegar to the dyebath. Pink lemonade can be made using a lemonade recipe and a red anthocyanin from fruit juice or a red tea. There are many other possibilities. The only limits are the students imaginations the restriction to use only purchased food products, and cooking pans, when dyeing food items.

Science Classes

Many experiments are suitable for physical science, chemistry and biology classes in middle school and high school. Many of these can illustrate the scientific method and use of appropriate controls. Others illustrate concepts of acid-base chemistry, redox chemistry, organic chemistry, physical chemistry and spectroscopy. Some of them will be detailed below along with suggestions for extending the activities if desired.

Anthocyanin Colors

An easy source of anthocyanins is herbal tea bags. A large variety of red or red orange teas containing hibiscus flowers will work. These are often sold as rose hip tea, orange tea, peach and rose tea and so forth. Read the ingredients and if you find rose hips and hibiscus they will produce some color changes. For younger students, cranberry juice can be used to avoid boiling water and will produce similar results. This activity can easily be used with students of all ages. Younger students simply add acids, bases and alum to the anthocyanin dyebaths and watch the color changes. Older students make the additions more carefully by performing qualitative titrations using acid, base and alum solutions and counting drops of added titrant. Chemistry students could perform the activities as quantitative titrations. Any students can extend the activity to include dyeing fabrics with the modified solutions.

Two version of the activity are presented here. Each includes prepared activity sheets for the students immediately following the description of the activity. The first activity is written to allow student to observe the color changes without need to accurately measure the additives. Sufficient amounts of vinegar, baking soda or alum are added to ensure a color change. This would be a fun activity for younger children and directions are included to use the household sources of the chemicals required. The second method is a qualitative titration where acid, base and alum solutions are made by adding measured volumes of the vinegar, alum and bicarbonate to distilled water. The dyebaths are qualitatively titrated by adding the solutions to the dyebath in 20 drop increments. The color changes will be slower and observations can be made after each addition. This is a good activity for older elementary or middle school students. There is some overlap on the recommended ages for these two activities as the science background of the students and the facilities available in the classroom will determine which experiment is most appropriate for any particular class. A method has not been provided for a quantitative titration, but a high school chemistry teacher can easily modify the activity to use standardized solutions of dilute hydrochloric acid, sodium hydroxide and aluminum salt to titrate the dye.

Both versions have an extension of the activity to dye fabrics with the various states of the anthocyanin dye. This can be used with either the tea dye or juice dye, however the solution will need to be heated at the onset of the dyeing process. The heating should be done by the teacher for younger children. The dyebaths are allowed to stand overnight and students can observe the resulting colors the next day. They will probably be surprised to see that the colors of the dyed fabrics will not always mirror the solution colors. Some of the solutions may change color overnight and students can record this observation as well. Both solutions and fabrics dyed with them are shown in color figure 5a and b. Note that the color of the fabric does not always mirror the color of the dyebath!

Version 1: Changing Colors of Fruits and Berries

This activity uses food products and is appropriate for even the youngest students. It is recommended for grades / years: pre-K – 6. The time required for set up, excluding shopping is 10 min plus time for tea to cool. If using cranberry juice, there is no cooling time required. Student groups of 4-6 children are suggested. The activity will take students about 15 min to perform without extension and about 30 min with extension depending on their age and comfort with working independently.

Supplies required per student group without extension

- 4 cups suitable for hot liquids (if using juice clear plastic cups can be substituted)
- 4 plastic spoons
- measuring spoons (½ tsp, 1 tsp. and 1 Tbs sizes)
- 1 or 2 L empty plastic pop bottle (optional)
- 4 "red" herbal tea bags (or 1 L of cranberry juice)
- 1 L of boiling water (if using tea bags- performed by teacher)
- 1 Tbs. vinegar
- 1 tsp. baking soda
- 2 tsp. pickling alum

Teacher Set-up for activity

Each student group will need 600 - 800 mL of tea or juice, depending on the size of the cups. They will need to fill all 4 cups about 2/3 full. The teacher can make sufficient tea for the entire group in one pot or pan using 4 herbal tea bags and 600-800 mL of water per student group. The tea should be steeped for 3-5 min and the tea bags removed. This can be done the day before the activity if desired. For safety reasons with young children, the tea should be made ahead and cooled to room temperature before the students use it. It is recommended that the cool tea for each group be transferred to a clean 1 or 2 L plastic pop bottle or pitcher to make it easy for student to pour the liquid themselves and to avoid one group spilling the tea for all the groups.

Additional supplies for the optional "fiber dyeing" extension of the activity

- 4 small (about 2" x 2") squares of pre-washed 100% bleached cotton fabric or 4 short lengths (4") of pre-washed white cotton or wool yarn.
- microwave oven (this step may be performed by the teacher)
- cafeteria tray to hold cups (optional but helpful)

Teacher set up for extension of activity

The teacher should allow an additional 15 min of set up for the extension excluding the time for machine washing and drying of the fabric or yarn. Fabric or yarn must be thoroughly washed according to the "easy scouring" directions in Ch. 2. Once washed and dried, the fabric should be cut into 2" X 2" squares. Alternately, cotton and/or wool yarn may be used instead. The yarn should be cut into 3-4" lengths. This can be done well ahead of the experiment and stored until needed.

Students will bring their fabric samples in their labeled dye cups to the teacher. Each cup should be heated in the microwave until hot but not boiling. (This usually takes about 2-3 min on low power setting per cup. It may be done after class time by the teacher or in class with the students present. The amount of time required will depend on the number of student groups). Once heated the samples should stand in the dyebaths overnight. Student should *not* be allowed to handle the cups with the hot solutions. Instead the teacher should put them onto a tray and set them aside until they have cooled. Students may make the first observations later in the day. Final observations can be made the next day and the fabric samples recovered, rinsed and dried.

A prepared student lab sheet is included on the next page. Teachers may make as many copies as required for their classes, however the sheets are copyright protected and cannot be published, displayed or presented except in the classroom.

Name________________

Changing Colors of Fruits and Berries

A student activity for grade/year pre-K – 6 taken from
The Science and Art of Natural Dyes © JMB 2006

Supplies required for your group

- 4 cups
- 4 plastic spoons
- measuring spoons (½ tsp, 1 tsp. and 1 Tbs sizes)
- 1 L of bottle of tea or juice prepared by your teacher
- 1 Tbs. vinegar
- 1 tsp. baking soda
- 2 tsp. alum

Introduction: Most red, pink and blue fruits, berries and flowers are colored by dyes called anthocyanins. Anthocyanins are natural chemicals made by plants. They change colors when they are mixed with other chemicals like vinegar, baking soda and alum. Isn't it amazing that one kind of chemical dye can make so many different colored flowers?

Procedure:

1. Get four cups and four plastic spoons. Mark each cup as A, B, V or A+B.

2. Get one bottle or pitcher of cooled red herbal tea or cranberry juice. (Your teacher will tell you which to use).

3. Fill each cup about 2/3 full with liquid.

4. Get cup V and add 1 tablespoon of vinegar to the cup and stir. What color is the dye?

5. Get cup B and add 1/2 teaspoon baking soda to the cup and stir. What color is the dye?

6. Get cup A and add 1 teaspoon of alum to the cup and stir. What color is the dye?

7. Get cup "A+B" and add 1/2 teaspoon baking soda. Slowly add 1 teaspoon of alum to the cup and stir. What happened when you added the alum?______What color is the dye?

8. Clean up following your teacher's directions.

Name ________________

The Science and Art of Teaching with Natural Dyes © JMB 2006
Dyeing Fabrics Extension

1. Place one fabric square in each of the cups and let it sit for about five minutes.

2. Use the spoon to lift the fabric out of one cup far enough to see it's color. Put it back in the cup and write the color of the fabric on the chart below. Do this for all four cups then bring the cups to your teacher. S/he will heat them because dyes work better at hotter temperatures. Let the dyes stand overnight.

3. The next morning dump the dye out of each cup. Let the fabric sit in the cup to dry. When the fabrics are dry put them on the chart below. Do one fabric at a time so you make sure to put it in the right place on the chart. Attach it with tape or glue.

4. Write your observations about how the fabric has changed since yesterday.

5. Write observations about whether the fabrics are the same color as the dyebath.

Name ______________________________ **The Science and Art of Teaching with Natural Dyes © JMB 2004**

Dyebath	Dyebath color Day 1	Fabric color Day 1	Dyebath color Day 2	Fabric sample Day 2
V + vinegar				
B + baking soda				
A + alum				
A+B + alum and baking soda				

Version 2: What factors affect anthocyanin colors?

This activity can be performed with food products for younger students or with laboratory grade chemicals for older students. It is recommended for grades / years: 5 - 9. The time required for set up, excluding shopping is 10 min. Student groups of 2-4 are suggested. The activity will take students about 30-45 min to perform without extension and about 30 additional min for the extension depending on their age and comfort with working independently.

Supplies required per student group without extension

- 4 250 mL Pyrex or Kimax beakers.
- 3 125 mL beakers

- 1 250 mL graduated cylinder
- 1 100 mL graduated cylinder
- 4 stirring rods
- measuring spoons (½ tsp, 1 tsp and 1 Tbs sizes)
- 3 droppers
- pH paper wide range or litmus paper
- 4 "red" herbal tea bags
- 1 L of boiling distilled water (may be provided by the teacher)
- 300 mL room temperatue distilled water
- 1 Tbs. distilled vinegar (or dilute acetic acid)
- 1 tsp. sodium bicarbonate (or baking soda)
- 2 tsp. alum (potassium aluminum sulfate or ammonium aluminum sulfate)

Teacher Set-up for activity

Each student group will need 200 mL of tea in each of the four beakers. The students or the teacher can also make one batch in a 1L beaker using 4 herbal tea bags and 900 mL of water per student group. The tea should be steeped for 3-5 min and the tea bags removed. For safety reasons, the tea should be made ahead and cooled to room temperature for younger students. This can be done the day before the activity if desired. Older students can prepare the tea themselves.

Additional supplies for the optional extension of the activity

- 8 small (about 2" x 2") squares of pre-washed 100% bleached cotton fabric or 4 short lengths (4") of white cotton or wool yarn.

- microwave oven, hotplate or Bunsen burner (this step may be performed by the teacher for younger students. See procedure for version 1 of this activity)

Teacher set up for extension of activity

The teacher should allow an additional 15 min of set up for the extension excluding the time for machine washing and drying of the fabric or yarn. Fabric or yarn must be thoroughly washed according to the "easy scouring" directions in Ch. 2. Once washed and dried, the fabric should be cut into 2" X 2" squares. Alternately, cotton and/or wool yarn may be used instead. The yarn should be cut into 3-4" lengths. This can be done well ahead of the experiment and stored until needed.

Older students will heat the labeled dye beakers themselves and younger students will bring them to the teacher for heating. They should be heated until hot but not boiling. (This usually takes about 2-3 min on low power in a microwave oven). It may be done after class time by the teacher or in class by older students. The beakers may also be heated using hot plates or Bunsen burners. Once heated the samples should be let stand overnight. The younger student should *not* be allowed to handle the cups with the hot solutions. Instead the teacher should put them onto a tray and set them aside until they have cooled. Students may make preliminary observations at that time or later in the day. In order to ensure adequate binding of dye to the fiber, the first fabric sample should not be recovered until the dyebath has cooled to room temperature. Final observations can be made the following day and the second fabric sample recovered.

Prepared student lab sheets for the activity and the extension are included on the following pages. Teachers may make as many copies as required for their classes. The sheets are copyright protected and cannot be published, displayed or presented except for use in the classroom

Name________________

What factors affect anthocyanin colors?

A student activity for grade/year 5 – 9 taken from
The Science and Art of Teaching with Natural Dyes © JMB 2006

Supplies

- 4 250 mL Pyrex or Kimax beakers.
- 3 125 mL beakers
- 1 250 mL graduated cylinder
- 1 100 mL graduated cylinder
- 4 stirring rods
- 3 droppers
- measuring spoons (½ tsp, 1 tsp and 1 Tbs sizes)
- pH paper, wide range or litmus paper
- 4 "red" herbal tea bags
- 1 L of boiling distilled water (may be provided by the teacher)
- 300 mL room temperatue distilled water
- 1 Tbs. distilled vinegar
- 1 tsp. sodium bicarbonate
- 2 tsp. alum (potassium aluminum sulfate)

Introduction: Most red, pink and blue fruits, berries and flowers are colored by pigments known as anthocyanins. Anthocyanins are natural chemicals made by plants. Anthocyanins undergo chemical changes when they are exposed to acids, bases and metal ions. In this experiment, you will examine the affect of acids (vinegar, a dilute solution of acetic acid), bases (sodium bicarbonate or baking soda) and a metal ion (aluminum in alum, potassium aluminum sulfate). Anthocyanins in plants also combine with other dye chemicals called co-pigments to produce a whole range of colors. Isn't it amazing to realize that one type of dye chemical makes nearly all the red, blue and purple flowers and fruits in the world?

Getting Ready

1. Get four 250 mL beakers and label each as acid titration, base titration, alum titration or alum+base titration. These will contain the tea dye you are making. They will be titrated with the substance named on the beaker.

2. Use a 250 mL graduated cylinder to add 200 mL of boiling distilled water to each of the beakers prepared in step 1. Place one anthocyanin tea bag in each beaker and let them steep for 3-5 minutes. It is important that all four beakers have the same amount of water and steep for the same amount of time. (Your teacher will tell you whether the water will be provided or if you have to boil the water yourself). After 3-5 minutes remove the tea bags from all four beakers and let them cool.

3. Get three 125 mL beakers to prepare the acid, base and alum solutions for the titrations. Mark the beakers as acetic acid stock, sodium bicarbonate stock and alum stock. Use a graduated cylinder to add 100 mL of distilled water to each beaker.

4. Prepare an acid solution adding two tablespoon of vinegar to the water in the beaker labeled "acetic acid stock." Stir the stock solution and set it aside.

5. Prepare a basic solution by adding 2 level teaspoons of sodium bicarbonate to the water in the beaker labeled "sodium bicarbonate stock." Stir the stock solution and set it aside.

6. Prepare an alum solution by adding 2 teaspoons of alum to the water in the beaker labeled "alum stock solution." Stir the stock and set it aside.

Performing the Titrations

7. Get the beaker labeled "acid titration" and use the pH paper to measure the pH. Record the initial color and initial pH on the data table. Next use a dropper to add 20 drops of the acetic acid stock to the beaker and stir. What color is the dye? Record it on the data table. Measure the pH and record that on the table as well. Continue to add the acetic acid stock solution in 20 drop increments until you have added a total of 100 drops. Don't forget to record the color and the pH after each 20 drop increment. Set both beakers and the dropper aside.

8. Get the beaker labeled "base titration" and use the pH paper to measure the pH. Record the initial pH and initial color on the data table. Next use a new dropper to add 20 drops of the sodium bicarbonate stock solution to the beaker and stir. What color is the dye? Record it on the data table. Measure the pH and record that on the table as well. Continue to add sodium bicarbonate stock solution in 20 drop increments until you have added a total of 100 drops. Don't forget to record the color and the pH after each 20 drop increment. Set both beakers and the dropper aside.

9. Get the beaker labeled "alum titration" and use the pH paper to measure the pH. Record the initial pH and initial color on the data table. Next use a dropper to add 20 drops of the alum stock solution to the cup and stir. What color is the dye? Record it on the data table. Measure the pH and record that on the table as well. Continue to add alum stock solution in 20 drop increments until you have added a total of 100 drops. Don't forget to record the color and check the pH after each 20 drop increment. Set the "alum titration" beaker aside. You will need to use the "alum stock solution and dropper for the next titration, so do not discard them.

10. Get the beaker labeled "alum+base" titration and add 1/2 teaspoon sodium bicarbonate to the cup and stir. You should be using the solid sodium bicarbonate, not the stock solution. Now, use the pH paper to measure the initial pH. Record the initial pH and initial color on the data table. Use the dropper to add 20 drops of alum stock solution to the cup and stir. What color is the dye? Record it on

the data table. Measure the pH and record that on the table as well. Continue to add alum stock solution in 20 drop increments until you have added a total of 100 drops. Don't forget to record the color and check the pH after each 20 drop increment.

Name ______________________________ The Science and Art of Teaching with Natural Dyes © JMB 2004

Data Table

	Vinegar (acetic acid) acid		Sodium bicarbonate alkalai/ base		Alum (potassium aluminum sulfate) Aluminum ion		Alum + bicarbonate Aluminum + base	
	color	pH	Color	pH	Color	pH	Color	pH
Initial value								
+20 drops								
+40 drops								
+60 drops								
+80 drops								
+100 drops								

Name ______________________________ The Science and Art of Teaching with Natural Dyes © JMB 2004

1. Place one fabric square in each of the cups and let it sit for about five minutes.
2. Use the spoon to lift the fabric out of one cup far enough to see it's color. Put it back in the cup and write the color of the fabric on the chart below. Do this for all four cups then bring the cups to your teacher. S/he will heat them because dyes work better at hotter temperatures. Let the dyes stand overnight.
3. The next morning dump the dye out of each cup. Let the fabric sit in the cup to dry. When the fabrics are dry put them on the chart below. Do one fabric at a time so you make sure to put it in the right place on the chart. Attach it with tape or glue.
4. Write your observations about how the fabric has changed since yesterday.
5. Write observations about whether the fabrics are the same color as the dyebath.

Dyebath	Dyebath color Day 1	Fabric color Day 1	Dyebath color Day 2	Fabric sample Day 2
V + vinegar				
B + baking soda				
A + alum				
A+B + alum and baking soda				

Other Extensions and Follow-Ups of this Activity for Science Students

There are a number of possible "follow up" activities of interest to biology, botany, biochemistry and/or chemistry students. Two are presented here, "True Blue?" and "Anthochlor Colors." These are both "exploratory" activities whose results will depend on the particular flowers, berries and other materials used. Stepwise procedures are not given and students should be encouraged to "experiment." More experienced science students should be able to do this, however they may be resistant to just "seeing what happens" rather than knowing what is "supposed to happen" before they begin.

True Blue?

As mentioned in the previous activity, anthocyanins are responsible for the colors of most red, blue and purple flowers. However, they are difficult to extract and use as a blue dye. The chemistry of anthocyanins and the reasons for the blue dye problems are explained in detail in Ch. 8. Students can collect or be given blue or violet flowers and nontoxic berries, such as blue cornflower, blue hydrangea, chicory, iris, blueberries and so forth. The whole flowers and berries can be treated with heat, acid vapors (vinegar or acetic acid will work well) to see if the color is affected. Biology students in particular may be interested in working with the intact flowers.

Alternately chemistry students may prefer to extract the dyes as in the previous activity ("What factors affect anthocyanin colors?") and monitor color changes over time as the solution is heated and subsequently cooled. The pH of the extract can be adjusted and the stability of the dye monitored. Students should find that anthocyanins are not stable at alkaline pH. In fact letting an alkaline anthocyanin dyebath stand overnight will result in the blue color becoming decidedly gray. Metal ions such as aluminum and iron can be added to the bath and the stability of these solutions monitored. In short, students can be asked to use the results of the previous lab activity to attempt to produce a useful blue dye from anthocyanin containing plants.

In the activity "What factors affect anthocyanin colors?" students should also have noticed the cloudiness when both alum and sodium bicarbonate were added to the clear red anthocyanin dyebath. Chemistry students should know that cloudiness often indicates a precipitate has formed. At the teacher's suggestion, the students may try to take advantage of this by precipitating the blue complex within the fiber or fabric. Students might simmer unmordanted fabric in an acidic red anthocyanin dye bath then add bicarbonate and alum to attempt to precipitate the complex within the fiber. Alternately simmering alum mordanted fabric in an acidic red dyebath and then adjusting the pH could be attempted. A number of different versions can be tried with the teacher providing as much or as little guidance as s/he chooses.

Alternately, students can be guided to attempt the precipitation by talking about natural dyes like indigo that precipitate within fibers before beginning the experiment. It is also helpful to introduce the historically important mineral dyes chrome yellow (lead chromate) and Prusian blue (iron ferrocyanide). Both dyes are precipitates formed within the fiber by metathesis reactions. The fabric is first soaked in a solution of soluble salt of the metal ion, usually lead nitrate and iron buff respectively, then immersed in a soluble solution of the anion, usually potassium chromate and potassium ferrocyanine respectively (Liles). The resulting metathesis results in the immediate formation of a precipitate within the fabric that remained trapped in its insoluble state.

The production of chrome yellow dye is dramatic. A properly trained science teacher may demonstrate it for the class in a test tube with small amounts of solution and fiber. Proper chemical hygiene should be practiced for this activity including the use of rubber gloves. Because of the high toxicity of dichromate, it is not recommended here as an activity for students. Under no circumstances should the students be allowed to handle the fiber after dyeing. After seeing the demonstrations, the students can them be encouraged to try something similar with the anthocyanin dyes.

Anthochlor colors

Anthochlors, like anthocyanins, undergo pH dependent color changes. In alkaline solutions, the normally yellow anthochlor pigments become orange or rosy orange. This color change is the result of ionization of one of the phenolic hydroxyl groups of the dye chemicals. The chemistry of anthochlors is described in detail in Ch 8.

As suggested for blue flowers, botany or biology students can explore this color change in the intact flowers by exposing them to ammonia vapors. The flowers should become reddish in the presence of the ammonia and revert to yellow when removed from the vapors. Many but not all yellow composites contain anthochlors and this color change can be demonstrated by comparing anthochlor containing flowers such as coreopsis, cosmos and dahlia with yellow flowers that don't have chalcones and aurones. The chemistry of anthochlors is discussed in more detail in Ch 8.

Chemistry and biochemistry students may be more interested in working with extracts than with the intact flowers. The extracts can be treated with base, as described above, but a more interesting application might be to monitor the uv/vis spectrum of the pigment at various pH's. The resulting change in electronic energy levels with ionization of the compound shifts the absorbance maximum (lambda max) and thus changes the reflected color. The relationship between absorbed wavelengths and the resultant color observed is explained in Ch 6. The ranges of maximum absorbance for chalcone and aurone anthochlors are listed in Ch 8.

Students may find the extracts contain too many chemicals to see the anthochlor spectra clearly and may choose to try different techniques to extract or purify the anthochlors for further study. This could be a good idea for a project oriented organic laboratory class or an "end of the year" project section of a traditional organic lab class. The students would be unlikely to find this project "written up" in a laboratory manual or chemical education journal and would be required to read and interpret the scientific literature in this field.

Alternately, students may be encouraged to find commercial sources for the pure compounds or study related phenol compounds to investigate the pH dependent spectral changes that occur. This would teach students how and why "model compounds" are often used to simplify the initial studies of natural products and natural systems. Purchasing authentic pure compounds or devising model studies might be more appropriate for students in analytical or physical chemistry classes.

Dyebath temperature

Nearly all the dye recipes in this book recommend simmering temperatures for both dye extraction and fiber dyeing. There are several reasons for this. First, the solubility of most dyestuffs is increased in hot water, thus the hotter the water, the more dye can be extracted. On the other hand, boiling temperatures will destroy most dye chemicals and result in reduced rather than increased color. As in many aspects of dyeing, simmering is a compromise to maximize yield and minimize damage.

Simmering is also recommended for the fiber dyeing process. Fabrics in hotter water are more swollen and open allowing dye to enter more readily. Dyes also need heat energy to overcome the potential energy barrier to penetrate the fibers as described in Ch 7. These are thermodynamic considerations. Migration of the dye to the fiber is also faster at higher temperatures. This is a kinetic consideration, but it is still important. A fiber will not be dyed well or evenly if the dye does not migrate to the fiber efficiently.

On the other hand, dye recipes often recommend leaving fibers in the dyebath overnight or until the bath cools. Other recipes suggest waiting a day after dyeing before rinsing the fiber. This is because the permanent binding of dyes to fibers is, in part, dependent on entropy considerations as described in Ch 7. Entropy binding effects are enhanced at lower temperatures, so it is best to let a dyed fiber cool before rinsing away dye that will be bound more securely at lower temperatures. A similar argument can be made for physical entrapment of dyes. Cotton in particular will return to a more tightly closed state when cooled and will physically trap dyes more effectively than when hot.

The activity suggested here allows students to observe the results of some of these factors. The time the fiber spends in the dyebath must be controlled in order to see the kinetic effects. Time is somewhat less important for the thermodynamic effects if the dyes are left to attain equilibrium, however in practice they will not have attained thermodynamic equilibrium when removed from the dyebath so all the samples should be in the dyebaths for exactly the same amount of time. The easiest way to do this is to prepare all the baths first then begin and end the dyeing at all temperatures simultaneously. The best results will be observed if the fibers to be used have been scoured, but mordanting is not necessary in most cases. The simplest and easiest dyes to prepare can be used for this experiment. Herbal tea dyes work particularly well, are quick to prepare, and being food products, are safe for even the youngest children.

The dyebaths must all be equally concentrated to control any effects caused by dye concentration differences. The best way to do this is to extract the dye in one batch and then divide it among several "dyebaths" for the temperature portion of the experiment. It can be diluted at this time if needed. The teacher can prepare the dyebath ahead of time, or students can prepare it one day and use it the next. It is best not to try to prepare and use the dyebath the same day because it will be difficult to cool it fast enough to test the lower temperatures. This experiment can be used with very young students if a teacher or teacher's aid controls the simmering and boiling dye baths. Alternately, the hotter baths can be omitted and hand-hot, warm, cool and cold baths used safely with even the youngest students. With older students, both a simmering and boiling bath are recommended along with a room temperature and an ice water bath. Additional dyebaths at temperatures between room temperature and simmering bath can be added if desired. Prepared lab sheets are provided.

Name ________________________

Too Hot, Too Cold and Just Right
An activity for grade/year 4-12

The Science and Art of Natural Dyes © JMB 2006

Time required: 40-60 min.

Purpose: To see how the temperature of the dyebath affects the dyeing of fibers.

Supplies per group (4 students suggested or 8 with younger groups)

- 4 250 mL graduated Pyrex or Kimax beakers or Styrofoam cups
- 4 stirring rods or plastic spoons
- 1 100 mL graduated cylinder or measuring cup
- 500 mL of dye stock (prepared by teacher)
- 4 fabric or fiber samples (2"x2" fabric or 4" length of yarn) pre-soaked in water
- Bunsen burner or hot plate if using boiling and simmering dyebaths
- Warm water, cool water and crushed ice or ice cubes (all distilled water if possible)

Procedure:

1. Get 4 beakers and label them "boiling," "simmering," "room temperature" and "ice cold." These are your dyebaths. If there are four in your group, each person should be in charge of one of the dyebaths. If there are more people, share the dyebaths.

2. Get 4 fabric samples and keep them in distilled water until ready to dye them.

3. Your teacher has prepared a concentrated dye stock for you. Use a graduated cylinder to add 100 mL of the dye stock to each beaker.

4. To the beakers labeled "boiling," "simmering" and "room temperature," add 100 mL of room temperature distilled water. Do not add anything to the "ice cold" beaker yet.

5. One person in your group should heat the beaker labeled "boiling" on a hot plate or Bunsen burner until it begins to boil. Keep the dyebath boiling. Add more hot water if you need it to keep the volume of the dyebath constant.

6. One person in your group should heat the beaker labeled "simmering" on a hot plate or Bunsen burner until bubbles form on the bottom of the beaker. Once the bubbles form, lower the heat. Try to keep bubbles on the bottom but do not let the dye boil!

7. While the beakers are heating, one person in your group should add the ice to the beaker labeled "ice cold." You must be careful to add 100 mL of ice to the beaker. If you are using crushed ice, measure it with a graduated cylinder. If you have ice cubes, the easiest way to measure the ice is to use the graduations on the beaker. Add ice to the beaker until the level of the dye stock goes up from the 100 mL mark to the 200 mL mark. (This is called measurement by water displacement).

8. Once all four dyebaths are ready, remove the fabric samples from the distilled water and squeeze them. Each person in your group should put one fabric sample in his/her dyebath. Be organized. You should all put the fabric in at exactly the same time then check the time. All the samples must stay in their dyebaths for the same amount of time. Usually this will be 20-30 minutes.

9. When the time is up, turn off the heat and immediately remove your fabric sample from your dyebath. Put it on a clean paper towel and let it sit until it reaches room temperature. When it's room temperature, rinse it in cool water and place on a new paper towel to dry. Keep track of which fabric sample is yours!

10. Tape your dry samples on the chart and comment on the results. Which dyebath produced the deepest color? What temperature works the best for dyeing fibers?

Bath temperature	Fabric Sample	Comments
Boiling		
Simmering		
Room Temperature		
Ice Cold		

Dyeing Different Fibers

Most of this book has focused on dyeing cotton and wool, but other natural fibers can be dyed with natural dyes. Even though these fibers are cellulosic, like cotton, and protein, like wool, differences in cellular structure, arrangement of the molecules, or amino acid composition of the proteins will result in different dyeing properties. Biology students can explore differences in cuticle structure in sheep wool compared to other hairs to understand why different keratin sources have different dyeing properties. They can compare silk, a non-keratin protein with the keratins in hair to understand why silk accepts dyes so differently from wool. Cellulosic fibers from different sources will have different dyeing properties depending on plant cell structure. Some detail about fibers is included in Ch. 7.

Even some synthetic fibers can be dyed with natural dyes, particularly nylon. Nylon accepts color surprisingly well with both natural and synthetic dyes intended for wool. This is because nylon is a polyamide, like the amino acid backbone of wool and other protein fibers. Polyesters, on the other hand, will not be colored as well by natural dyes. Students learning about polymers and polymer chemistry can explore the similarities and differences among the fibers and their ability to accept natural dyes. More information on natural and synthetic fibers can be found in the references and in college text books.

The experiment written here calls for individual fabrics to be purchased and used. Multifiber test strips are also available. These strips have bands of different fiber wefts woven together on one fabric. They are more expensive than purchasing fabrics individually however they come "ready to use" so are less tedious for the teacher to prepare. They have the added advantage of being impossible for students to mislabel or mix up, which occurs surprisingly often when using separate fabric samples. The multifiber test strips can be obtained from at least two sources. Shirley Developments Ltd., Machine Control B.A.A. CANANDA INC., 701 Ave. Meloche, Dorval, Quebec H9P 2S4, Canada (Epp) and The Society of Dyers and Colourists (Trotman).

I generally use individual fabrics because they allow the greatest flexibility of fiber types and sample size. Students can also be involved in the choices of the fibers and natural fibers can be emphasized more than would be the case with the multifiber strips. Several yards can be purchased at a time and scoured at once, then used for years. At times I have produced my own multifiber strips by cutting and sewing the strips of various fabrics together, then cutting them into lengths of mixed fibers, however this can be quite tedious. Most often I rely on the students to keep their samples straight. I provide a set of undyed "reference" fabrics at the front of the room to help the groups that inevitably mix up their fabrics. A permanent felt tip fabric marker can also be supplied for the students to label the fabrics before dyeing to prevent misidentification.

The "Material Matters" activity is quite flexible and can be used in a number of different classes and class settings from primary school through college non-majors chemistry and biology courses. The number of fibers and dyes can be adjusted by the teacher depending on the class size and age of the students to ensure success. Easy "tea bag" dyes or more difficult plant dyes can be used depending on the class. It can be used as a "fun" first or last lab for college science majors in special topics, biochemistry or botany courses on check-in or check-out day. I have found that the lab runs more smoothly if each group is responsible for one dye. The groups can be quite large with younger students, and much smaller with older students. Each group dyes samples of each fiber for every student. This minimizes the number of students handling the boiling dyebaths and minimizes the potential for dye mix ups and mislabeling. It also ensures that each student sees the entire process of dyeing rather than spending the lab period wandering from dyebath to dyebath while a partner does all the work. It can be a good exercise in cooperative learning for younger students because each group must depend on all the other groups to get their dyed samples.

Material Matters

Age grade/year 5- college

Time required: 90 min – 2 hr.

The time required for this activity will vary considerably with the experience of the class. There are two periods of simmering required, for 30 min. each. The remainder of the time suggested is for organizing and clean up. More experienced science students may complete the activity in as little as 80-90 min.

Supplies for the class

- 2"x2" sample of fabrics. One square for each student.

Suggested are cotton, wool, silk, linen, ramie, nylon and polyester.

- 4" length of fiber. One length for each student in the class.

Suggested are jute, raffia, acrylic yarn.

- Other fibers/substances if desired

Suggested are purchased feathers, dried reeds or rushes, toothpicks or popsicle sticks, leather, hard boiled eggs (must **not** be eaten after dyeing).

Supplies per group:

- 1 4-6 L Pyrex or Kimax beaker
- 1-2 2 L Pyrex or Kimax beaker
- 2 stirring rods
- one old nylon "knee high" stockings or a wire mesh or strainer
- Hot plate or Bunsen burner and ring stand
- Distilled water
- Dyestuff
- Fiber samples. (One sample of each type of fiber for each student in your class).

Material Matters
A student activity for grade 5-college from

The Science and Art of Natural Dyes © JMB 2006

Purpose: To see how well natural dyes work on a variety of natural and synthetic fibers.

Supplies for the class

- 2"x2" sample of fabrics. One square for each student.

Suggested are cotton, wool, silk, linen, ramie, nylon and polyester.

- 4" length of fiber. One length for each student in the class.

Suggested are jute, raffia, acrylic yarn.

- Other fibers/substances (optional)

Possibilities include purchased feathers, reeds, rushes, wood, leather, egg shells

Supplies per group:

- 1 4-6 L Pyrex or Kimax beaker
- 1-2 2 L Pyrex or Kimax beaker(s)
- 2 stirring rods
- one old nylon "knee high" stocking or a wire mesh or wire strainer
- Hot plate or Bunsen burner and ring stand
- Distilled water
- Dyestuff
- Fiber samples. (One sample of each type of fiber for each student in your class).

Procedure:

1. If you are using nylon "knee highs", place the dyestuff your group has chosen into the knee high and knot the top. Put the stocking into the larger beaker and add water to cover it. You may need to push it down with a stirring rod to keep it submerged. If you are not using knee highs, place the dyestuff into the beaker and add sufficient water to cover it.

2. Heat the dyebath with a hotplate or Bunsen burner to nearly boiling. Turn down

the heat and simmer for 30 min, stirring occasionally. If you are using a "knee high" to hold the dyestuff, you should stir more frequently.

3. While the bath is simmering, get the fiber samples. Remember you need one of each type of fiber for every person in your class. Place them in a 2L beaker and add warm water to cover all the fibers and fabric samples. Use a stirring rod to push them to the bottom from time to time.

4. After the dye has simmered for 30 minutes, remove the dyestuff. If your dyestuff was contained in a knee high, simply remove the stocking, squeeze the excess liquid out as you would with a tea bag (remember it's hot!). Discard the stocking and contents. If you simmered loose dyestuff instead, you will have to strain the dyebath. The best way to do this is to with a wire strainer or wire mesh. Pour the dyebath through the mesh into a second 2L beaker. Use the new beaker as your dyebath. Return it to the heat source.

5. Remove the fibers from the water where they are soaking and gently squeeze them. Carefully add them to the dyebath and simmer for an additional 30 minutes. Do not let the dyebath boil with the fibers in it. If it begins to boil, shut off the Bunsen burner or remove the beaker from the hotplate for a few minutes.

6. If possible let the samples stand in the dyebath until cool. If this is not possible, remove the fibers carefully and let them cool for a few minutes. Once cool rinse them in running water and spread them on paper towels to dry.

7. Organize the fibers so that each type of fiber is in a pile. Include information about the dye used and the type of fiber next to the pile.

8. Once all groups have completed the dyeing process, collect your set of samples. You should have one sample of each fiber dyed in every dye.

9. Arrange the samples onto a sheet of paper or into your lab notebook as directed by your teacher or lab instructor. Your instructor will tell you whether to answer the questions individually or if they will be discussed by the class.

Questions:

In general, do natural or synthetic fibers dye better with natural dyes?

Which natural fibers were dyed the deepest color? The palest?

Which synthetic fibers were dyed the deepest color? The palest?

Which non-fabrics were dyed the deepest color? The palest?

Are all animal fibers dyed to the same extent? Discuss why you think this is so.

Are all plant fibers dyed to the same extent? Discuss why you think this is so.

What did you find to be the most surprising result of this activity? (For example, a fiber you didn't think would dye well but did, or a fiber you expected would dye well that did not, and so forth).

Mordanting

Most natural dyes require the use of a mordant to produce a fast dye, however some substantive dyes need no mordant. Conducting a mordanting experiment can be a good illustration of the Scientific Method. It clearly demonstrates the importance of using an appropriate control before drawing conclusions. A mordanting experiment is also a great way to illustrate the chemistry of mordants and dye binding. If the dyes are carefully chosen, students comparing a pre-mordanted fiber with an unmordanted fiber will see three or four different results. Most often the pre-mordanted fiber will be dyed the same color but a deeper shade when compared to the unmordanted fiber. Sometimes however, there will be little or no discernable difference between mordanted and unmordanted fibers. Other times the fiber will take up little or no color at all without a mordant. And occasionally the mordanted fiber may be dyed a different color than the unmordanted sample.

All these results can be explained by understanding why dyes are colored and how dyes bind to fabrics. In general the role of a mordant is to serve as a bridge between the dye and the fiber as explained in Ch. 7. The mordant forms a more secure connection between fiber and dye resulting in more dye being retained by the fiber. Some dyes will have no affinity at all for the fiber without a mordant. These will not be colored at all. Most dyes will have reduced affinity and thus paler colors with the unmordanted fibers. Conversely, some substantive dyes will show little or no difference with or without a mordant because the dyestuffs contain mordant along with dye. The fiber is simultaneously mordanted and dyed as described in Ch 3 & 7. Finally color changes with mordanted fibers occur when the dye-metal complexes have different electronic energy levels than the uncomplexed dye. Changes in energy levels change the wavelengths of light absorbed and reflected by the dyed fiber as explained in Ch. 6.

These experiments can be used to illustrate the use of Scientific Method alone, or they can be used to teach dye chemistry. The science teacher can talk about dye chemistry to the students and then let them do the experiment to illustrate the principles they have discussed. Alternately, students can "discover" the differences themselves to serve as motivation to explore the chemistry of dyes in more detail. Of course the level of chemistry can easily be adjusted for the age and background of the students and the level of the course.

"Alum or No Alum" and "Mordant Montage" Activities

Careful selection of the dyes used in this activity is critical if students are to see the various effects of mordants. A number of different dyes can be used and will produce good results. At least three different dyes are required. A maximum of six is recommended as an abundance of dyes will not illustrate the principles any better but is more likely to result in confusion of the samples.

To see the different results discussed above, at least one dye must be from the "mordant independent" category, labeled "type A" below. This category includes only substantive dyes that require no external mordants to produce color. However not every substantive dye will be independent of external mordants. At least one dye should be from the other extreme, producing little or no color without external mordant. These are all mordant or adjective dyes and are grouped as "type C" below. Once again, not all adjective dyes will fall into this category.

The third dye should color the fabric with or without a mordant, but produce a deeper shade or a different color on mordanted and unmordanted fabric. This category will include both the substantive and adjective dyes. Here the category is called "mordant affected dyes" to encompass both traditional dye categories. Suitable dyes are listed as type B below. Any of the dyes in the table may be chosen provided at least one dye from each category is used. The dyes presented here are those easy to obtain and prepare, or dyes that show the effects particularly well. Other dyes from Ch. 3-5 may be used if desired.

Dyestuffs used to extract natural dyes are all natural products and will vary in color and fastness from batch to batch. If time allows, teachers should pre-test the activity with several dyes to see which will be best for their students. This is particularly true if teachers choose dyes not listed below. Alternately, students can be given dyes in categories labeled "A", "B" and "C" and asked to choose dyes themselves, one from each category. In this case it is suggested that different student groups use different dyes and share results to maximize the number of dyes students can compare.

Type A, substantive dyes, show little or no difference between mordanted and unmordanted fibers. They include Barberry *wood* dye Ch 3, Black walnut *brown* Ch 4, Generic eucalypt dye Ch 3, Generic bark dye Ch 3, and Tea bag dye Ch 5. Type B, mordant affected dyes, show a noticeable difference between mordanted and unmordanted fibers. Mordanted fibers may be a deeper or different color. They include Coreopsis dye Ch 4, Goldenrod dye Ch 4, Marigold dye Ch 3, Mint tea bag dye Ch 5, Onion dye Ch 3 and Weld yellow dye Ch 5. Finally Type C, mordant dyes show a very large difference between mordanted and unmordanted fibers. Unmordanted fibers usually retain little or no color. These dyes include Barberry *leaf* yellow Ch 3, Cranberry dye Ch 5, Mulberry dye, Ch 5, Rosy herbal tea dye Ch 5, and Chamomile yellow Ch 5

This activity can be used with any age student if sufficient care is taken to protect young children from scalding. The teacher should adjust the level of difficulty depending on the background and age of the students. Two versions of the activity are presented here. The first is a truncated version with fewer variables and is suggested for younger and/or less experienced students of upper elementary school. Either wool fabric or yarn may be used in this version. Cotton is not suggested because of the additional mordanting required. Three specific dyes are suggested for this version of the activity as younger students will probably fare better with fewer samples. Specific dyes are suggested because they are easy to obtain and use with children and should produce reliable differences. Other dyes from table 10.2 may be substituted provided one dyestuff from each column is used. Teachers using this truncated activity for older students may choose to provide either more dyes or both cotton and wool fabrics for their classes. In that case, the cotton fabric should be mordanted using the tannin/alum mordanting recipe. This activity requires fiber to be mordanted by the teacher in advance of the activity and saved for later use.

Alum or No Alum

"Alum or No Alum" can be performed by even younger students, K-3, if the entire class functions as a "group." The teacher can assign subgroups for each dye and either mordanted or unmordanted yarn. The subgroups can label their cups and put the dyestuff into the cups then bring them to the teacher. The cups can be placed on a tray and the teacher can add boiling water to all the cups at the same time. When the dyes are ready for use the students can bring the yarn sample to the teacher and the teacher can put the sample into the hot water. It is not safe for young children to work with hot or boiling water. The teacher must emphasize the danger of scalding to the students and place the cups out of reach of the curious students while the dyes steep and while the yarn samples are dyed. After the dye baths are cool, the teacher can safely let each dye group reclaim their dyebath for observation and comparison with other students. Alternately, teacher aids or volunteer parent classroom helpers can be recruited to handle the boiling water for each group.

Alum or No Alum Suggested grade/year 4-8

Supplies required per group (6 students suggested)

- alum mordanted wool fabric (3"x3") or alum mordanted wool yarn (6" length)
- unmordanted wool fabric (3"x3") or unmordanted wool yarn (6" length)
- two tea bags (or other type "A" dye) or dyebath prepared by teacher
- handful of dry onion skins (or other type "B" dye) or dyebath prepared by teacher
- two chamomile tea bags (or other type "C" dye) or dyebath prepared by teacher
- 6 Styrofoam cups, coffee mugs or Pyrex beakers
- 6 plastic spoons
- paper towels
- scissors
- cafeteria tray (optional but helpful)

Teacher set-up:

This activity requires the use of pre-mordanted fibers. They may be prepared by the teacher for class use or prepared ahead of time by the students if time permits. With young children, the teacher may choose to prepare the dyebaths ahead of time as well.

Name _______________

Alum or No Alum

A student activity for grade/year 4-8 taken from

The Science and Art of Natural Dyes © JMB 2006

Procedure:

1. Each student in your group should get one cup and write his or her name on the cup.

2. Each student in your group should get either one tea bag or half the onion skins and put it into his/her cup. Write "T" on your cup if you have a regular tea bag. Write "O" on your cup if you have onion skins. Write "C" on your cup if you have the chamomile. Make sure you write down the name of the dye you get.

3. When everyone in your group is ready, tell your teacher that you are ready for the boiling water. CAUTION: Boiling water can cause burns! Your teacher will ask you to bring your cups to the front or to place them all on a tray on your desk. The cups will be your dyebaths and need to steep for about 5 minutes.

4. While you are waiting for your dyebaths to be ready you need to organize your group into the "alums" and the "no alums." Split into two subgroups so that each subgroup has one cup "T," one cup "O" and one cup "C." Once this is organized, write your name on the "alum" or "no alum" side of the worksheet next to your dye. Now you are ready to get your yarn.

5. The "alums" subgroup should get alum mordanted yarn. Your teacher will have these set out on a paper towel or tray with **ALUM** written on it. Remember one student with cup "T", one with cup "O" and one with cup "C" should get **ALUM** mordanted yarn. Make sure you remember that you are an "alum."

6. The "no alums" subgroup should get unmordanted yarn. Your teacher will have these set out on a paper towel or tray with **NO ALUM** written on it. Remember one student with cup "T", one with cup "O" and one with cup "C" should get the **NO ALUM** yarn. Make sure you remember that you are a "no alum."

7. Now you are ready to dye your yarn. Your group must work together to do this right. Your teacher may want to supervise this step. When you are ready, count to three and each of you drop your yarn into your dye and stir it up.

8. Your teacher will ask you to set your dyebaths aside or will come to collect them so you can look at them later.

9. When your teacher tells you to look at your dyed fabrics, remove the fabric, rinse it in cold water and put it on a paper towel to dry. Make sure you keep track of

your own sample. When it is dry, cut it into 6 pieces, one for each person in your group. The other group members should also give you a sample of their fabric. Tape or glue all the samples to the proper place on the activity sheet and compare the results.

Name ____

Alum or No Alum Lab Sheet

The Alums	The No Alums
Put the name of the "no alum" student working with the dye next to the dye name. Tea: Onion: Chamomile:	Put the name of the "no alum" student working with the dye next to the dye name. Tea: Onion: Chamomile:

The Alum Fabrics	Dye Name	The No Alum Fabrics
	Tea	
	Onion	
	Chamomile	

Mordant Montage Suggested grades 9- college

This longer version of the mordanting activity, "Mordant Montage," is intended for older students and those with more science background. In the activity as written, students will be required to boil and handle boiling water. The ages suggested are only a guideline. Each teacher must be sure his or her students are mature enough to do this safely. If there is any doubt, the activity should be run with modifications similar to those used for preceding "Alum or No Alum" activity.

In the Mordant Montage activity, both cotton and wool are tested. Three different mordanting methods are compared on each fabric. Suggested mordants are the traditional wool mordant alum/cream of tartar, the traditional cotton mordant, alum/tannin and no mordant. It is suggested that six dyes be used, two from each category in table 10.2. This activity requires significant organization by the class and groups and is most appropriate for classes with good laboratory technique and lots of experience.

For younger or less experienced students, teachers may want to modify the experiment to decrease the complexity. For example, each group of six students could work with only one dye. Each student in the group would then handle only one type of fiber and one dye. After dyeing is complete, the samples from all the groups could be compared to answer the questions and draw conclusions. This would work well in a collaborative learning model.

Teachers wishing to increase the level of complexity for college science students might include additional mordant choices such as alum alone and/or tannin alone. If the teacher and students are properly trained in handling heavy metals, then other traditional mordants could be used to increase the color variation. The properly trained science teacher is directed to Ch. 2 and the references for information on mordants other than alum. Alternately, the activity could be extended further by having students test additional fabrics or fibers such as linen, raffia, silk and so forth. Time required will vary depending on how the activity is run (see procedure) but is estimated at 1.5-3 hours.

Name ____________________

Mordant Montage
A student activity for grade/year 9-college taken from

The Science and Art of Natural Dyes © JMB 2006

Supplies per group (6+ students) or per class if each group works with one dye.

- 6 Alum/cream of tartar mordanted wool fabric (3"x3") or yarn (6" length) sample
- 6 Alum/cream or tartar mordanted cotton fabric (3"x3") or yarn (6" length) sample
- 6 Alum/tannin mordanted wool fabric (3"x3") or yarn (6" length) sample
- 6 Alum/tannin mordanted cotton fabric (3"x3") or yarn (6" length) sample
- 6 unmordanted wool fabric (3"x3") or yarn (6" length) sample
- 6 unmordanted cotton fabric (3"x3") or yarn (6" length) sample
- two different type "A" dyes or dyebaths prepared by teacher
- two different type "B" dyes or dyebaths prepared by teacher
- two different type "C" dyes or dyebaths prepared by teacher
- 6 pyrex beakers
- 6 stirring rods
- paper towels
- scissors
- hot plates or Bunsen burners and ring stands
- permanent felt tip marker
- wire strainer or mesh if students are preparing the dyebaths

Procedure:

1. Label 6 beakers with the name of the dye and your name or initials.

2. Each student in the group should be responsible for one dye. Your teacher will have all the dyes labeled as type A, type B or type C dyes. Each group needs two different "A" dyes, two different "B" dyes and two different "C" dyes, so work together when choosing your dyes.

3. Each student will also need 6 fabric or fiber samples. You should have alum/cream of tartar mordanted wool and cotton fabrics or fibers, alum/tannin mordanted wool and cotton fabrics or fibers and unmordanted fabric or fiber. It is critical that you keep these straight, so use the permanent marker to write on each fabric "a/ct", "a/t" or "no m" and whether it is wool or cotton fabric. Soak all the fabrics and fibers in hot water until ready for use. Keep the mordanted and unmordanted fibers separate.

4. When your group is organized, you are ready to begin. Your teacher will tell you if your class will boil the dyestuffs to prepare the dyes, or if s/he prepared them for you ahead of time. Either way, the dyebaths must be simmered. Boiling water can cause burns so use care.

5. If your teacher prepared the dyebaths for you skip to step 7. If you are preparing the dyebaths put your dyestuff into the beaker and add distilled water to cover it. Your teacher will give you the amounts of dyestuff and water to add or will give you a recipe sheet. Heat until nearly boiling then reduce the heat and simmer for about 30 min.

6. Remove the dye stuff from the dyebath by straining through a wire mesh or strainer and return the bath to the heat. Skip to step 8.

7. Heat the dyebath prepared by your teacher until simmering (bubbles appear on the bottom but do not break the surface) and keep it simmering.

8. It is important that all the fabrics be simmered for the same amount of time, and that you know which is which. Make sure they are labeled with a permanent marker before you put them into the dyebaths! Remove the fabrics from the soaking bath and squeeze them gently. Add them all to the dyebath at the same time. Simmer for 30 minutes then remove the fabric samples. When they have cooled to room temperature, rinse and dry them.

9. Arrange the samples on a sheet of paper and compare them.

<u>Some questions to think about</u>:

Do the wool and cotton fibers show differences with the same mordant and dye?

How do the mordanted fibers compare with the unmordanted?

Does one mordanting method appear to be better than another?

Is the same method good for both fabrics?

<u>Testing Dye Fastness</u>

With High School and College Students

Testing light and color-fastness of dyes is an excellent way to bring "real life" chemistry applications into the classroom. Tests can be as simple or complex as the teacher chooses. Students could dye fabrics with a number of different natural dyes and then test them for light-fastness, wash-fastness and perspiration fastness as part of an integrated unit. Comparing fabrics dyed with natural dyes, with dyes purchased for use at home and with commercially dyed fabrics could be an interesting application for a science class.

Light-fastness in particular involves some interesting and complex radical photochemistry, described in Ch. 9, that would be of interest to chemistry students. The photoxidation and photoreduction of dyes would make interesting "real life" examples for students studying photochemistry in college chemistry courses. This would be an excellent opportunity to bring in a guest speaker or seminar speaker from industry who could talk about the photochemistry of dyes. It is also a good way to force reluctant students to read the chemical literature. There are many examples of dye photochemistry in journals but very few in their text books.

Although journal articles will be primarily studies with synthetic dyes, the chemistry is relevant to natural dyes. Many examples can be found in journals such as *Dyes and Pigments, Textile Research Journal* and *Journal of the Society of Dyers and Colourists.* Since most colleges and universities will not have these journals in their own libraries, such a project can also serve as an introduction to the use of the interlibrary loan or the electronic journal center. Electronic resources are becoming increasingly common and undoubtedly will continue to do so in the future. Both services are available to Ohio college and university students on-line through Ohiolink.

With Younger Students

Even the youngest students can learn chemistry from testing dyes for light-fastness. Rather than complex explanations, children can simply be told that the energy from the sunlight breaks bonds in the chemicals and thus destroys the dye. Young children can understand bonded chemicals as atoms "holding hands" to make a chain. Playing a game like "Red Rover" can help to illustrate that it takes a lot of energy to break the bonds (running works, walking doesn't) and so wavelengths of sunlight with more energy will break more bonds. Red Rover can also illustrate the concept that some bonds break more easily than others, Nearly all the children will know that some parts of the Red Rover line are easier to break through than others!

After trying one of two of the activities that follow, students could be encouraged to devise tests for other types of fastness, for example chlorinated swimming pool water followed by sun exposure. Other possibilities are the effect of wash temperature, or type of laundry detergent on fading. Most parents will have noticed differences in fading of clothing depending on the laundry detergent used. Indeed there are even television commercials that address laundry fading which many students will have seen. Students could interview parents about laundry fading as part of an integrated curriculum. Guest speakers from the detergent industry could also be invited to visit the classroom.

Light-Fastness

Testing dyes for light-fastness in industry requires comparison of the dyes tested to a set of standard dyes, as described in Ch 9. Rigorous protocols must be followed and specific equipment is needed. A simpler method is described by Wingate in which the dyes are characterized by the number of "standard fading hours" of light exposure before appreciable fading is noted. This could easily be used with students.

In most cases, however, students can benefit by following a similar but simplified procedure. No standards or special equipment are needed for the activity presented here, but the method follows the industrial procedure in other respects. Students compare color changes of a dye to a control sample of the same dye. Rather than assigning a number to describe the light-fastness, the dye is rated subjectively as having "good," "moderate," or "poor" light-fastness as compared to the other samples.

Although natural dyes are notoriously fugitive, they do not fade as quickly as the students might imagine. Fading may take weeks (Pyott, Wingate) so maximizing sun exposure is essential. Dyes should be chosen carefully to include some with poor light-fastness. Anthocyanin tea dyes are a good choice and others can be chosen from reading Ch 3-5. The experiment will be most relevant to the students if the dyes used are those they have produced in other experiments or for other projects. Testing for fastness would be a great follow-up activity to a dye day or traditional folk craft day.

This activity allows students to construct the light fading apparatus from plate glass, cardstock and foam-core board. For young students, teachers should construct and handle the apparatus to minimize the risk of injury from glass breakage. Students must follow relevant safety practices when working with glass, such as the use of safety glasses at all times. Replacement of the glass with plastic is not recommended as most plastic does not transmit uv light well and will hinder the fading.

Light-Fast or Fugitive? All ages (with modifications for younger students)

Supplies required per group (2-6 recommended)

- 1 light fading apparatus constructed by the teacher OR

1 piece plate glass (approx. 8"x10")
1 piece of foam core board or particle board (approx 8"x10")
3 sheets heavy cardstock or cardboard
masking tape
4 scraps foam core (about 1" sq.) or rubber stoppers or other spacers

- 3-4 2"X2" pieces of fabric dyed with different natural dyes
- masking tape
- straight pins
- scissors

Teacher set-up: Additional time is required if the teacher builds the light boxes. Minimal time is required to set up the activity aside from collecting the required materials.

Light-Fast or Fugitive?

An activity for testing the light-fastness of natural dyes taken from
The Science and Art of Natural Dyes © JMB 2006

Supplies required per group (2-6 recommended)

- 1 light fading apparatus constructed by the teacher OR

1 piece plate glass (approx. 8"x10")
1 piece of foam core board or particle board (approx 8"x10")
white glue
colored pencil or thin-line permanent marker
4 scraps foam core (about 1" sq.) or foam rubber spacers

- 3-4 2"X6" pieces of fabric dyed with different natural dyes
- 1 sheets of cardstock or heavy cardboard (approx. 8"x10")
- masking tape
- straight pins
- scissors

Procedure

Part 1: Building the light testing apparatus.

1. If your teacher has done this for you, skip to part 2. To build the apparatus you will need the supplies listed above by the dot symbols. The apparatus will look like a sandwich, with the foam core sheet on the bottom, one spacer in each of the corners and the plate glass on the top. The dyes to be tested will go into the middle.

2. First put the foam core sheet on the table with the long dimension up and down, the way you would put a sheet of paper. Now use a ruler and a pencil or permanent marker draw lines to divide the foam core from top to bottom into three approximately equal columns. Try to avoid making dents in the foam core when you do this and make the lines as straight and parallel as you can. At the very top, number the columns 1-3 from left to right. Measure the height and width of columns 1 and 2 and write it down.

3. Cut four squares of about 1" from the foam core scraps and glue one to each corner of the foam core sheet with white glue. These will be spacers to hold the

glass away from the fabric. Let the apparatus stand until the glue is completely dry.

4. While the glue is drying, cut two cardstock strips. Cut one strip the same width as column 2 of the foam core but about one inch shorter. Use a straight pin or thumbtack to make a small hole in each corner and write "2" on the strip so you can identify it later. Cut the second piece of cardstock the same height as column 1 but about one inch wider and write "1" on it. Put holes in the corners with the pin or tack.

5. Put the cardstock strips and glass aside until the glue on the foam core base is dry and you have finished part 2.

Part 2: Arranging the fabrics.

7. Once the glue is dry on the foam core you can begin arranging the fabrics. Take the first fabric to be tested and place it on the foam core with its long length going across the short dimension of the foam core. Don't cover the column numbers. The fabric should overlap all three columns and should have about the same amount of fabric in columns one and three. Once the fabric is arranged correctly, use straight pins to secure the four corners of the fabrics to the foam core. Add more pins to the top and bottom edge of the fabric at each of the column lines. Push the pins in at an angle if necessary to sink them most of the way into the foam core. Do this for all the fabrics to be tested.

8. After all the fabrics are arranged, place the cardstock labeled "2" over column 2 and make sure it covers the middle section of all the fabrics. When it is in the right place, secure the corners with straight pins.

9. Now cover the first column with the cardstock labled "1." This cardstock should overlap the column 2 cardstock by about 1 inch. Once it is right, secure the corners with straight pins. Now use masking tape to cover the seam where the cardstocks overlap to make sure no light can get in. You will need to remove the tape part way through the experiment, so don't use too much.

10. When both pieces of cardstock have been secured, place the plate glass sheet on top of the sandwich. Wrap masking tape around the top and bottom edges to hold the glass to the foam core so you can move it without breaking the glass. Don't cover all the edges with tape. It is important that air can freely circulate.

Part 3: Testing for light-fastness

12. You need to place the light testing apparatus where it will get as much sun as possible throughout the day. There should not be any shadows on the box during the day. Your teacher will give you some suggestions. There may be a safe place outside for the boxes, but it will probably be a south facing window in your

classroom (in the southern hemisphere it should be a north facing window). The box needs to be propped up at an angle about equal to your latitude. For Ohio and nearby areas, an angle of about 40 degrees will be close enough. A music stand or bookstand is a good support for your box.

13. Let the apparatus sit in the window one week of full sun then remove it from the window. Carefully open the glass and remove the cardstock covering column 1. Replace the glass and return the apparatus to the window for another week of full sun. If there are a lot of rainy days, you may need to let it sit for more than two weeks to complete the experiment.

14. Remove the apparatus from the window and open it. Remove the cardstock from column 2 and compare the color with the portions that have been exposed to sun. If there is a noticeable difference then remove the fabrics. If there has been little fading, cover column two again, reassemble the box and replace in the window for another week or more.

Wash-Fastness

Tests for fastness to washing and other factors do not require external standards as light-fastness testing does. Instead the degree of fading is determined by a comparison with a gray scale as described in Ch 9. Students can follow the procedures for temperature and detergent type and wash their fabrics following the International Organization for Standardization, ISO, protocols with only slight modifications. This would be a good industrial application of chemistry to use with high school and college students. The protocols are detailed and stringent controls are maintained. This will be more appropriate for some classes than others and the teacher may choose to use the more rigorous protocols or to use the modified methods suggested here.

Younger or less experienced science students in particular, would probably find it easier and more interesting to instead compare different brands of laundry detergent that they have at home or have seen advertised. This activity is no less relevant to "real life" applications and would be a very good activity for "Chemistry for Consumers" types of courses taught in high school and college. Most students in these classes will not go on to careers in the chemical industry. Some will have careers with little or no chemistry involved, but all of them will have to do laundry.

Fastness testing could easily be incorporated in an integrated curriculum as well. There are many possible ways to do this. Students could "produce" and "market" their own detergent. As part of the "product development" wash testing would be necessary. There are published procedures for making soaps and detergents with children (see for example the Partners for Terrific Science series). Any of these recipes could be used to prepare a soap and the soap used in the test for wash-fastness. Alternately, students could develop and market their own naturally dyed fabrics. The fastness of the dye would be part of the information that consumers would need. History is also relevant here. As mentioned earlier re-dyeing faded items was common practice in the past when fast dyes were not available. Even most early synthetic dyes were not very fast.

The key to making a wash-fastness activity work for a variety of ages is the amount of detail used in the washing method. One version is provided where older, more experienced students can carry out the washing according to ISO standard methods in a washing machine. A second method is provided where younger students can simply compare several different detergents when washed by hand. In this activity, it is helpful to compare brands with and without bleach, and "bargain" brands to those advertising superior stain removal properties. The later usually have additives not always found in the bargain brands.

Wash-Fastness

An activity for grades/years 10- college taken from
The Science and Art of Natural Dyes © JMB 2006

Purpose: To test the wash-fastness of natural dyes using industrial ISO methods. Both dye fading and bleeding will be tested simultaneously.

Supplies per group

- Cotton or wool fabric sample(s) dyed with natural dye (about 18 cm x18 cm square)
- One undyed 12 cm x 12 cm square of cotton
- One undyed 12 cm x 12cm square of wool
- Laundry detergent
- Washing soda (sodium carbonate)
- Permanent fabric marker
- Thermometers
- Drying oven or clothes dryer
- Washing machine access (optional but very helpful)
- Sewing machine access or hand sewing supplies (needles, thread etc.)

Procedure:

1. Cut the dyed sample into four 4 cm x 10 cm strips so that you have one piece for each ISO method and one to serve as an unwashed "control." Three of the five ISO methods are presented here, however your group might not be evaluating all three methods. Check with your teacher if you are not sure.

2. Use the permanent marker to label each piece of fabric with the name of dye and fiber and the ISO method that will be used to wash it.

3. Get the undyed cotton and wool and cut both into 4 cm x 10 cm strips.

4. Sandwich each fabric to be tested between an undyed strip of wool and an undyed strip of cotton. Sew around all four sides of the fabric sandwich. Save the fourth piece of each fabric as a control or reference sample.

5. Run the ISO methods as described below (Pyott, Trotman). All 3 methods use 5g soap per liter water. Method 1 uses 40 C (104 F) water, and a 30 min cycle. Method 2 requires 50 C (122 F) water and a 45 min cycle. Method 3 requires the addition

of 2 g anhydrous sodium carbonate (washing soda) per liter with 60 C (140 F) water and a 30 min cycle.

6. Rinse the samples two times in cold water (in the washing machine) and then rinse again for 10 minutes in running tap water.

7. After washing and before drying, remove the sewing on three sides **only** of the sample and then dry the fabric in an oven at 60 C/140 F.

8. Compare the dry fabric to the control fabric to see if fading has occurred. Cut a small sample of each to attach to the data sheet.

9. Compare the undyed cotton and wool coverings to control cotton and wool to see if bleeding has occurred. Cut off a small sample to attach to the data sheet.

10. Record the information on a data table below or in your notebook. Be sure to compare all three ISO methods. If your teacher has provided an ISO grayscale, use it to assign a fastness rating to the dye.

	ISO 1 method	ISO 2 method	ISO 3 method
Control dyed sample			
Laundered dyed sample			
Control white cotton sample			
Tested white cotton sample			
Control white wool sample			
Tested white wool sample			

Name ____________________

Does the Dye Come Out in the Wash?

An activity for grades/years K-9 taken from
The Science and Art of Natural Dyes © JMB 2006

Purpose: To see how natural dyes stand up to different laundry detergents

Supplies per group:

- 2 4 inch squares of fabric dyed with the same natural dye
- Several brands of laundry detergent
- Water (distilled preferred but tap water can substitute)
- Bucket or dishpan
- Drying rack or clothes line

Procedure:

1. Get your fabric samples.
2. One fabric is the "control." Glue or tape it to your data sheet.
3. The other fabric sample is your "test sample."
4. Get a bucket or dishpan and the laundry detergent for your group.
5. Add a little bit of detergent to the bucket.
6. Fill the bucket ¾ full with warm water. It should feel like bath water.
7. Put the "test sample" fabric into the water and swish it around for 5 minutes.
8. Empty the wash water into the sink and rinse out the bucket.
9. Fill the bucket with clean water.
10. Put the soapy sample into the bucket and swish it around.
11. Empty the rinse water. Rinse the fabric in running water at the sink.
12. Let it dry and tape or glue it to the data sheet.
13. Compare the test sample and the control to see if the test sample faded.
14. Compare your sample to samples washed with different detergents by the other groups.

Name ____

Does the Dye Come Out in the Wash?

Data Sheet
An activity for grades/years K-9 taken from
The Science and Art of Natural Dyes © JMB 2004

Write the name of the laundry detergent that you used	
Control fabric sample	**(tape sample here)**
Washed fabric	**(tape sample here)**
Did it fade?	
Which detergent caused the least fading?	
Which detergent caused the most fading?	

Other Fastness Tests

Fastness to perspiration is an area of much research. Recipes for "artificial perspiration" can be found in many references (Aranyosi *et. al.* 1998 and 1999, Csepregi *et. al.* 1998b, Trotman). Different industry standards are used in different countries. Students can compare color-fastness of natural dyes exposed to artificial perspiration to that of natural dyes. Once again, a gray scale is used to evaluate the results.

Swimwear needs to be tested for fastness to chlorinated water. The activity suggested here asks students to devise their own test for swimwear. They must work out a protocol and must be sure to include factors such as time in the water, water temperature and subsequent exposure to the sunlight. Once students have devised the protocol they will use it to test their dyes. Students will find this activity easier to do if they have previously performed one of the other fastness tests.

Still more tests are done for evaluating fastness to rubbing or crocking, fastness to sea water, and so forth. Additionally, a number or tests are done to evaluate the fastness of dyes to industrial processing steps, such as Mercerization (Trotman)

Teacher Set-Up for Perspiration Fastness

Most of the work should be done by the students. The teacher simply needs to make sure the supplies the students will need are available in the lab or stockroom. Some chemicals may have to be ordered ahead of time and placed in the lab or stockroom for student access. This activity is primarily designed for A/P high school or college students. Other high school students with accelerated science backgrounds may be able to perform the activity with help from the teacher.

Supplies required per group

- access to electronic journals or interlibrary loan.
- lab oven at 37°C
- naturally dyed cotton or wool fabric (8 cm X 10 cm)
- scoured white cotton (8 cm X 10 cm)
- scoured white wool (8 cm X 10 cm)
- 2 glass plates
- 10 lb weight that can be placed in the oven
- distilled water.
- Access to stockroom chemicals. Recipes vary, but very common are:
- sodium chloride
- organic acids (acetic, lactic)
- organic salts (sodium panthotenate),
- an orthophosphate salt (Na2HPO4• 12 H20)
- amino acids (his, asp)
- a monosaccharide (glucose)

Perspiration Fastness

From The Science and Art of Natural Dyes © JMB 2006
Supplies required (per group)

- access to electronic journals or interlibrary loan.
- lab oven at 37°C
- naturally dyed cotton or wool fabric (8 cm X 10 cm)
- scoured white cotton (8 cm X 10 cm)
- scoured white wool (8 cm X 10 cm)
- 2 glass plates
- 10 lb weight that can be placed in the oven
- distilled water.
- access to stockroom chemicals.

Procedure:

1. Search the literature to find a references for artificial perspiration. Electronic journals such as *Dyes and Pigments* will probably be your best source.

2. Prepare the artificial perspiration as described in the literature. You will need a solution that is 50 times the weight of your fiber sample including the undyed sandwiching fabrics (see below).

3. Cut two 4 cm x 10 cm strips from the dyed fabric. One strip will serve as your test sample. The second strip should be saved as a "control" or "reference."

4. Get the undyed cotton and wool and cut each into two 4 cm x 10 cm strips. One strip of cotton and one strip of wool will be used for the test. The other samples will be controls of references. .

5. Sandwich the test sample dyed fabric between an undyed strip of wool and an undyed strip of cotton. Sew around all four sides of the fabric sandwich.

6. If a method is given in your reference, you should follow that to test the samples. Otherwise follow the standard method given below.

7. Immerse the fabric sandwich in 50 times its weight of artificial perspiration solution and let stand for 30 min at room temperature.

8. Decant the liquid then place the fabric between two glass plates. Place the plates in the oven and weight it down with the 10 lb weight. Let stand in the oven for 4 hr.

9. Evaluate the results and prepare a report. The report should include samples of the control and test fabrics. Be sure to look for and comment on fading and/or discoloration of the dyed sample as well as staining of the adjacent white fabrics.

Teacher set up for Swimwear fastness

This activity can easily be adjusted to a wide range of student abilities and ages. In all cases, the students are given responsibility for writing the protocol and conducting the tests and the teacher will serve as a facilitator. The teacher will provide some guidelines to the students at the beginning of the activity. The guidelines will vary depending on the age and experience level of the students. For example, younger students should *not* be allowed to use the concentrated pool chemicals. Instead they might use water collected from a public or private swimming pool for their test. Alternately, they might find the criteria and ask the teacher to prepare any required solutions. Conversely older students might well need to be told that they *cannot* simply collect water from the town swimming pool, but need to use test solutions of known composition and concentration.

As mentioned above, this activity is suitable for middle school students depending on the protocol used. Younger students should be encouraged to use real pool water for the test. Students may choose to dry the samples in sunlight using a modification of the light-fastness apparatus. Older student in advanced high school classes or college classes should use a more controlled protocol. They should use an "artificial pool water" formulation as is done in perspiration testing. The protocol should include the concentrations of the various "pool chemicals" used in the "artificial pool water" solution. Similarly the drying should be done under controlled temperature and sunlight conditions.

Student Procedure:

Swimwear needs to be tested for fastness to chlorinated water. This activity asks your class to devise your own test for swimwear. You must work out a protocol. Be sure to include factors such as time in the water, water temperature and subsequent exposure to the sunlight. Once you have devised the protocol you will use it to test your natural dyes. This activity will be easier to do if you have previously performed one of the other fastness tests.

You should turn in:

- a copy of your protocol
- a control sample of your fabric
- a test sample of your fabric
- a report on your findings

Other Activities

Teachers can add to the activities suggested here. The possibilities are nearly endless. There are some additional references with activities involving, dyes, fibers and fabrics, and detergent testing suggested in the Teacher Bibliography and Book Review section that follows. The activities in these references can be used to supplement the dye activities here in the design of a curriculum focusing on fibers and fabrics.

Chapter 11
Teacher Bibliography and Book Review

The complete references used in researching and writing this book are provided, however most teachers will not want or need those references. If I have been successful in my goal, most of you will find all the information you need in this book. The bibliography provided here is for those teachers who find the area fascinating enough to do further study.

In most cases, the books are referred to by author name only. The complete title and additional publishing information for these books, including the ISBN number, can be found in the Complete List of References at the end of this monograph. In cases where I recommend a book that I have not cited as a reference, ordering information is given with the description of the book. This bibliography provides a brief review of each book, specifically the age level of student activities, the target audience of the book and the amount of science content. This will assist teachers in finding books most appropriate for their interests and teaching levels.

Student Activity Books

These include books with additional activities on dyes, fabrics and fibers or cleaning products and detergents that could be used to supplement the activities provided here.

Epp, Dianne Vat Dyes and Natural Dyes are short books of activities. Each book contains about 5 activities and includes simplified chemistry and background for each activity. Teacher materials such as overhead templates and student activity sheets are provided. Activities are most appropriate for middle school or introductory high school chemistry classes. Historical and cultural information is included for some activities to help the teacher integrate the activity with other student learning. These are relatively inexpensive books and some teachers will want to purchase them.

Pyott, Steven Textile Care and the Consumer. This book is designed for home economics/ consumer science students at a college level. There are a number of easy activities including testing of soaps and detergents to determine their ingredients, and tests to determine fiber content of fabrics. Black and white pictures of gray scales and light- and wash-fast tested samples are included in the text. Discussion of testing for fastness to light, washing, water, rubbing (crocking) is included but only general suggestions are given for related activities rather than activity protocols for fastness. Most science teachers will probably prefer to borrow this book from the library.

The "Science in Our World Teacher Modules" set includes modules on soap, cleaning products and polymers. Their format is similar to the books by Epp. They include a range of activities appropriate for various ages from primary school to high school. The typical target audience age/level varies somewhat from module to module. Those with fiber and detergent activities are Polymer Chemistry and Institutional Products Chemistry. Both books are available from Terrific Science Press as are the books by Epp. Contact the Center for Chemical Education, Miami University Middletown, Middletown OH, e-mail: CCE@muohio.edu.

Teacher References

This section includes list of dye books with varying amounts of dye chemistry. It is not intended to be an inclusive list, but rather a completely biased list of my favorite books. Teachers interested in natural dyes who like this book will probably enjoy these books as well. The 'science books' by Cannon & Cannon and Liles are my very favorite and most frequently used references when I teach science students about natural dyes. The others 'hobbyist' books on the list have less science but are well written and reliable books about natural dyeing. They are the books I turn to when I am looking for dye recipes, dyeing for fun or just want to read about natural dyes from a less scientific point of view. I own copies of all the books on this list and recommend them highly. These are not the only dye books I own, but are the books I find myself turning to most often.

Science Books

Cannon, John and Margaret Dye Plants and Dyeing is a wonderful reference book about wool dyeing with natural dyes. The Cannons are both trained botanists and Margaret is now a weaver specializing in dyeing and spinning. This book provides correct botanical information, lovely botanical illustrations and dye colors produced for a number of historically important dyes from around the world. It does not include specific recipes but does give general guidelines at the beginning of the book. This is one of my favorite dye books and most useful references when teaching natural dyes.

Liles, J.N. The Art and Craft of Natural Dyeing is a valuable, scientifically accurate reference book about dyeing wool, cotton, linen and silk with color plates of samples and detailed recipes for a number of historically important dyes. Liles is a emeritus professor of zoology at the University of Tennessee and approaches dyeing as a scientific exercise. His recipes are detailed, well tested and properly referenced. This book is one of my favorites and very useful when teaching natural dyes.

Hobbyist Books

Androsko, Rita Natural Dyes and Home Dyeing is now available as a reprint of an older book. As might be expected, it is black and white and thus not particularly attractive, but it has many detailed recipes. Androsko is scientifically literate and is careful and accurate in her use of science. She lists a helpful appendix of the antiquated chemical names (as does Liles) found in old dye recipes with their more systematic current names. It is relatively inexpensive and is a useful reference to own.

Buchanan, Rita A Dyer's Garden is a small, affordable paperback book designed for the dyer wanting to grow natural dye plants at home. Buchanan is an editor at Interweave Press (a publisher specializing in fiber craft books). She served as an editor for the well known Brooklyn Botanical Garden "Dyes from Nature" book. The Dyer's Garden focuses on plants easy to grow in the U.S. and includes recipes. The science is more limited than in Cannon & Cannon but what is provided is accurate. Unlike Cannon & Cannon, photographs rather than drawings are included for the plants and the dyed samples. Both cotton and wool are included in many recipes.

Lesch, Alma Vegetable Dyeing like Androsko is an older but reliable book. There are some color photographs in the original. It has tested recipes and accurate information. If you can find it through a used book dealer for a reasonable price, it is a useful book to own. Like Androsko, Lesch was using natural dyes during a period of time when most people had abandoned them in favor of synthetic dyes.

van Stralen, Trudy Indigo, Madder and Marigold is an exquisite dye book featuring mostly imported dyestuffs to dye wool. VanStralen teaches dyeing workshops and dyes large amounts of wool for her own weaving business. Once again, the science is more limited but what is included is accurate. Beautiful photographs of the dyeing process, the dyes and dyed fibers are presented. This would make a good "coffee table" dye book because the photographs are so attractive.

Specialized Books

Bryan, Nonabah G. and Young, Stella Navajo Native Dyes is now available as a Dover reprint of a 1940 book. Young, a home economist and Bryan, a Navajo weaver collaborated to document the traditional recipes used by the Navajo as well as some of Bryan's own invention. Fascinating book and well worth owning if you are interested in Navajo or Native American culture.

Carman, Jean K. Dyemaking with Eucalypts was published in 1978 but is difficult to find. It was published in Australia and Australian books seem to go out of print much faster than books published in America. If you have access to eucalyptus leaves, it is well worth trying to find a copy of this book. Carman dyes both wool and cotton with eucalypt dyes and includes recipes and color photographs.

Colton, Mary-Russell Ferrell Hopi Dyes. was originally published in 1965. Some used copies are available through used book dealers. It was scheduled to be issued as a Dover reprint so may be more readily available. As the name suggests, it includes a number of traditional Hopi dyes. It is particularly interesting because the Hopi traditionally use cotton rather than wool fibers.

Chemical Education Journal Articles

The articles are arranged in this section by the type of the article. They are presented in chronological order from newest to oldest references . A table is provided at the end of the section to divide the articles by dye topic.

I. Background Information Only- no demonstrations or activities

Boykin, David W. *J. Chem. Educ.* 75 (1998) 769. "A Convenient Apparatus for Small Scale Dyeing with Indigo"
Describes an apparatus made from filtering flask that can be used to dye with indigo. The apparatus minimizes the odor and amount of reducing agent needed for indigo dyeing. Most appropriate for college class or possibly well-equipped high school labs with advanced chemistry courses.

Levey, Martin *J Chem. Educ.* 32 (1955) 625 "Dyes and Dyeing in Ancient Mesopotamia"
Historical article describing the ancient use of indigo blue, kermes red, turmeric and saffron yellows and murex purples.

Sequin-Frey, M. *J. Chem. Educ.* 58 (1981) 301.
A brief description of some historically important natural dyes. The article focuses mainly on the more well-known dyes, particularly those from India and Europe. It includes the structural formulae and chemical categories of the dyes.

Stallmann, O. *J Chem. Educ.* 37 (1960) 220. "The Use of Metal Complexes"
Presents a brief historical introduction of metal mordants with natural dyes but mostly focuses on synthetic dyes. Does discuss bond formation between fiber and dye and the role of metal ions.

Steinhart, Carol J *J. Chem. Educ.* 78 (2001) 1444 "Biology of the Blues: The Snail Behind the Ancient Dyes"
A brief description of three species of Muricidae family snails used in the ancient Mediterranean dye industry. The article includes the chemistry of the reaction from the colorless gland secretion to the purple dye. It does not mention the dye used in Pre-Columbian America, although other references say it is the same dye from related snail species.

II. Background Information Including Teacher Demonstrations

Solomon, Sally, Hur, Chinhyu *J. Chem. Educ.* 72 (1995) 730. "Overhead Projector Spectrum of Polymethine Dyes"
Synthetic cyanine dyes are used with a home-made spectrometer. The "spectrometer" is constructed using an overhead projector, a diffraction grating and cardboard shield with two slits, although specific details for construction of the spectrometer leave much to the reader. The absorbance band of the dye is observed as a black band in the visible spectrum

projected onto the overhead screen. While presented as a demonstration for college physical chemistry classes, it would be even more appropriate for high school or junior high school teachers without access to commercial spectrophotometers. Although presented for cyanine dyes, it could be used with other dyes with sufficiently strong absorbance in the visible region.

Uzelmeier, Calvin E. III and Breyer, Arthur C. *J. Chem. Educ.* 75 (1998) 183. "Red Shoe, Blue Shoe"

This clever demonstration utilizes the acid-base color change of the synthetic dye Congo Red. A woman's dyeable shoe is dyed with Congo Red and placed in acid and base solutions to alternately change the color. The demonstation is made more dramatic by using food coloring to color the acid and base solutions so that the shoe turns red when placed in a blue solution and blue when placed in a red solution. The demonstration could be used as such to show the acid-base dependence on dye colors in general or could perhaps be adapted for use with natural dyes, such as anthocyanins. This could be used with any level of students.

III. Student Experiments with Varying Amounts of Background Information

Bahnick, Donald A *J. Chem. Educ.* 71 (1994) 171. "The Use of Hückel Molecular Orbital Theory in Visible Spectra of Polymethine Dyes"

Presents an alternative to the FEMO method for analyzing the correlation between structure and color in cyanine dyes. Most appropriate for a college physical chemistry or advanced organic chemistry class.

Cutright, R., Rynerson, J.A., and Markwell, J. *J. Chem. Educ.* 71 (1994) 682 "Fruit Anthocyanins" Colorful Sensors of Molecular Millieu"

Describes anthocyanin structure and shows absorbance spectra, including pH effects, of a number of anthocyanin containing commercial products. Helpful tables detail the specific anthocyanins found in the various products. Good experiment for college classes or high school AP classes with access to a good spectrophotometer.

Cutright, R., Rynerson, J.A., and Markwell, J. *J. Chem. Educ.* 73 (1996) 306 "Anthocyanins: Model Compounds for More than pH"

Good teacher background including the confusing nomenclature of anthocyanins and the anthocyanin structures. Students hydrolyze the anthocyanin glycones to the aglycones and analyze them by paper chromatography. This lab would be appropriate for high school or college students.

daQuesta, Celeste; Queirós, M.A.; Rodrigues, Ligia M. *J. Chem. Educ.* 78 (2001) 236. "Determination of Flavonoids in Wine by High Performance Liquid Chromatography"

The experiment includes the hydrolysis of the flavonoid glycones in wine followed by reverse phase HPLC to identify the flavonoids. The hydrolysis is carried out in one four-hour lab period and the HPLC in a second. Standards are run by the teacher for comparison of retention times and students identify and quantitate the flavonols in their wine sample. Most appropriate for a college organic or analytical chemistry lab.

Epp, Diane *J. Chem. Educ.* 72 (1995) "A World of Color: The Chemistry of Vat Dyes"

This is a version of the inkodye activity included in her Vat Dye book. It is appropriate for various levels from elementary through high school.

Fernelius, W. Conrad and Renfrew, Edgar E. *J. Chem. Educ.* 60 (1983) 633. "Indigo"

Brief history of indigo and dye chemistry are included. The experiment is a synthesis of indigo intended for college organic chemistry students.

McKone, Harold T. *J. Chem. Educ.* 56 (1979) 676 "Rapid Isolation of Carotenoids from Tomato Paste, Carrot Puree and Citrus Fruit Skin"

Process for isolation of the carotenoids lycopene and beta-carotene. The method is fast and easy, and does not require special equipment. It is most appropriate for college because of the use of organic solvents. It could be done in high school AP courses if there are laboratories equipped for the safe use of organic solvents.

Mebane,Robert C. and Rybolt, Thomas R. *J. Chem. Educ.* 64 (1987) 291 "Chemistry in the Dyeing of Eggs"

Very good background presented on the chemistry of dyeing eggs, including the role of the protein cuticle and the chemical structure of food dyes. The experiment allows students to access the importance of pH, and the presence or absence of the egg's cuticle in order to deduce the mechanism of dye binding to eggs. Very well organized and through experiment appropriate for junior high school through college audiences to illustrate dye principles, egg dyeing or simply the use of the scientific method to solve problems.

Silveira, Augustine Jr. and Evans, Jeffrey M. *J. Chem. Educ.* 72 (1995) 374 "Flash Chromatographic Separation and Electronic Absorbance Spectra of Carotenoids"

The separation is accomplished with a silica column packed using compressed air with 2-5 psi of pressure. The spectra require a good spectrophotometer. This would be most appropriate for a college organic or analytical chemistry lab course.

Smestad, Greg P and Gratzel, Michael *J. Chem. Educ.* 75 (1998) 752 "A Natural Dye-Sensitizer Energy Converter"

Student build an energy converter that utilizes a dye to collect light energy. The experiment requires student background from several science disciplines and is designed for an integrated science class. It is most appropriate for college or advanced high school students.

Soltzberg, Leonard J. *J. Chem. Educ.* 78 (2001) 1432 "Electronic Spectra of Conjugated Systems: A Modern Update of a Classic Experiment"

Alternative computer molecular modeling methods to replace the FEMO method of modeling of cyanine dyes. Appropriate for college physical chemistry classes.

Suzucki, Chieko *J. Chem. Educ.* 68 (1991) 588 "Making Colorful Patterns on Paper Dyed with Red Cabbage"

Background of the red cabbage dye that is a combination of anthocyanins and flavonols. Includes a discussion of the pH chemistry of anthocyanins and flavonols. The activity involves dyeing and then folding the dyed paper. Corners of the folded paper are dipped into solutions of differing pH to cause color changes. The papers are unfolded and dried and have colorful patterns produced by the pH dependence of the dye colors. The folding method and color dyeing is a modification of a traditional Japanese dyeing technique so this fits well into an integrated curriculum studying Asia or Japan. The activity is safe for elementary students but would be interesting for higher levels as well.

Torimoto, Narboru *J Chem. Educ.* 64 (1987) 332. "An Indigo Plant as a Teaching Material"

This intriguing activity uses fresh and dried indigo leaves for a variety of dyeing projects. The fresh leaves are applied using a flower pounding technique to produce colored cloth. Both fresh and dried leaves are used to dye fabric. The plants must be grown by the teacher or class to harvest the leaves. The article includes a brief description of indigo chemistry. Appropriate for junior high to college chemistry. The pounding of leaves could be done with elementary school students.

Table 11.1 Journal articles arranged by topic and level. For ease in reading the table, only the surname of the first author and year of publication are listed here. The complete references, brief descriptions and details of suggested uses with various age students are given above. The Roman Numeral in parenthesis after the author refers to list I,II or III above.

	Course Level			
Activity Topic	Elementary	Junior High/ Middle School	High School	College
Anthocyanins	Suzucki, 1991 (III)	Suzucki, 1991 (III)	Cutright, 1996 (III); Cutright, 1994 (III), Suzucki, 1991 (III)	Cutright, 1996 (III); Cutright, 1994 (III), Suzucki, 1991; (III)
Carotenoids		McKone, 1979 (III)	McKone, 1979 (III)	Silveira, 1995 (III); McKone, 1979 (III)
Flavonoids	Suzucki, 1991 (III)	Suzucki, 1991 (III)	Suzucki, 1991 (III)	daQuesta, 2001 (III): Suzucki, 1991 (III)
Indigoids/ Vat Dyes	Epp, 1995 (III); Torimoto 1987 (III)	Steinhart, 2001 (I) Epp, 1995 (III); Torimoto 1987 (III)	Steinhart, 2001 (I) Epp, 1995 (III); Torimoto 1987 (III)	Steinhart, 2001 (I); Boykin, 1998 (I); Torimoto 1987 (III)
Synthetic/Cyanine Dyes	Uzelmeier, 1998 (II);	Uzelmeier, 1998 (II), Solomon, 1995 (II)	Stallman, 1960 (I), Uzelmeier, 1998 (II); Solomon, 1995 (II)	Stallman, 1960 (I) Uzelmeier, 1998 (II); Solomon, 1995 (II); Soltzberg, 2001 (III); Bahnick, 1994 (III)
General Interest	Levey, 1955 (I); Mebane, 1997 (III)	Levey, 1955 (I); Solomon, 1995 (II); Mebane, 1997 (III)	Sequin-Frey, 1981 (I); Stallman, 1960 (I); Levey, 1955 (I); Solomon, 1995 (II); Smestad, 1998 (III), Mebane, 1997; (III)	Sequin-Frey, 1981 (I); Stallman, 1960 (I), Levey, 1955 (I); Solomon, 1995 (II) Smestad, 1998 (III); Mebane, 1997 (III)

Complete List of References

Androsko, Rita J. [1968] 1971 *Natural Dyes and Home Dyeing"* New York: Dover reprint. ISBN: 0-486-22688-3.

Aranyosi, P, Csepregi, Zs, Rusznák, I, Töke, L. and Vig, A, *Dyes and Pigments* 37 (1998) 33-45.

Aranyosi, P, Czilik, M, Remi, E, Parlagh, G, Vig, A and Rusznák, I, , *Dyes and Pigments* 43 (1999) 173-82.

Aspland, J. R. 1998 "Colorants: Dyes" in *Color for Science, Art and Technology*, Kurt Nassau ed. Amsterdam: Elsevier ISBN:0-444-89846-8.

Bassett, Lynne Z. and Larkin, Jack 1998 *Northern Comforts: New England's Early Quilts 1780-1850.* Nashville TN: Rutledge Hill Press.

Bide, Martin 2000 "Secrets of the Printers' Palette: Colors and Dyes in Rhode Island" in *Down by the Old Mill Stream: Quilts in Rhode Island,* Linda Welters and Margarent T. Ordoñez eds. Kent OH: Kent State University Press.

Boehm, Bruce A. 1982 "The Minor Flavonoids" in *The Flavonoids: Advances in Research,* J.B. Harborne and T.J. Marbry eds. London: Chapman and Hall ISBN: 0-412-22480-1.

Brackman, Barbara 1989 *Clues in the Calico: A Guide to Identifying and Dating Antique Quilts.* McLean VA: EPM Publications ISBN: 0-939009-27-7

Bronson, J. & R. [1817] 1977. *Early American Weaving and Dyeing: The Domestic Manufacturers' Assistant and Family Directory in the Arts of Weaving and Dyeing.* New York: Dover reprint ISBN: 0-486-23440-1.

Brunello, Franco [1973] 1978 *The Art of Dyeing in the History of Mankind.* translated by Bernard Hickey, Cleveland OH: Phoenix Dye Works.

Bryan, Nonabah G. and Young, Stella [1940] 2002. *Navajo Native Dyes: Their Preparation and Use.* New York:Dover reprint. ISBN:0-486-42105-8.

Buchanan, Rita 1995 *A Dyers Garden: From Plant to Pot Growing Dyes for Natural Fibers.* Loveland CO: Interweave Press. ISBN:1-883010-07-1.

Burnett, J.H. 1976 "Function of Carotenoids Other than in Photosynthesis" in *Chemistry and Biolchemistry of Plant Pigments vol. 1, 2nd ed.* T. W. Goodwin ed. London: Academic Press.

Butterworth, Jeffrey A. 2000 "Quercitron: A North American Dyestuff" in *Down by the Old Mill Stream: Quilts in Rhode Island,* Linda Welters and Margarent T. Ordoñez, eds. Kent OH: Kent State University Press.

Cannon, John and Cannon, Margaret 1994 *Dye Plants and Dyeing.* Portland OR: Timber Press. ISBN: 0-88192-302-8.

Carman, Jean K. 1978 *Dyemaking with Eucalypts* Adelaide SA Australia: Rigby Ltd.

Carman, Jean 1984 "The Eucalypt Dyes" and "The Extension of the Eucalypt Dyes" in *Dyeing for Fibre and Fabrics* Janet deBoer, ed. St. Lucia Queensland Australia: U. of Queensland ISBN: 0-959455-1.

Casselman, Karen Leigh 1993 *Craft of the Dyer: Colour from Plants and Lichens.* New York: Dover

Castino, Ruth A. [1974]1975 *Spinning and Dyeing the Natural Way* London: Evans Brothers Ltd. ISBN:0-237-44817-3.

Cheesman, Patricia 1984 "Indigo Textiles, Japan, Laos, Nigeria: Indigo in Laos" in *Dyeing for*

Fibre and Fabrics Janet deBoer, ed. St. Lucia Queensland Australia: U. of Queensland ISBN: 0-959455-1.

Cheynier, V. 1999 "Tannins in Grapes" in *Tannins in Livestock and Human Nutrition. ACIAR Proceedings No. 92* J.D. Brooker, ed. Canberra: Australian Centre for International Agricultural Research.

Christie, R.M. 2001 *Colour Chemistry* Cambridge England: The Royal Society of Chemistry, ISBN:0-85404-573-2.

Colton, Mary-Russell Ferrell 1965. *Hopi Dyes*.Flagstaff AZ USA: Museum of Northern Arizona Press. ISBN:0-89734-000-0.

Cribb, A.B. and Cribb, T.W. 1981 *Useful Wild Plants in Australia* Sydney: Collins ISBN: 0-00-216441-8.

Csepregi, Zs, Aranyosi, P, Rusznák, I, Töke, L, Frankl, J and Vig, A. *Dyes and Pigments* 37 (1998) 1-14. (referenced as 1998a).

Csepregi, Zs, Aranyosi, P, Rusznák, I, Töke, L, and Vig, A. *Dyes and Pigments* 37 (1998) 15-31. (referenced as 1998b).

Cutright, R., Rynerson, J.A., and Markwell, J. *J. Chem. Educ.* 73 (1996) 306-08.

deBoer, Janet 1984 "Kasuri and the Art of Sunthetic Indigo Dyeing" in *Dyeing for Fibre and Fabrics* Janet deBoer, ed. St. Lucia Queensland Australia: U. of Queensland ISBN: 0-959455-1.

Devon T.K and Scott, A.I. 1975 *Handbook of Naturally Occurring Compounds, vol. 1* New York: Academic Press.

Epp, Diane N 1995 *The Chemistry of Natural Dyes* Middletown OH USA: Terrific Science Press ISBN:1-883822-06-8.

Finley, Ruth E. [1929] 1992 *Old Patchwork Quilts and the Women Who Made Them.* McLean VA: EPM Publications reprint. ISBN: 0-939-00968-4.

Frye, K., Ed. 1981 *The Encyclopedia of Minerology* Stroudsburg PA: Hutchinson Ross Publishing co. pp. 62-66, 474-80, 533-749.

Gallagher, Michael 1990 in *Flowers of the Loom: Plants, Dyes and Oriental Rugs*, Ian Close ed. Sydney: Royal Botanical Gardens. ISBN:0-7305-7471-7.

Gerber, F.H. 1984 "Indigo-Science and Art" in *Dyeing for Fibre and Fabrics* Janet deBoer, ed. St. Lucia Queensland Australia: U. of Queensland ISBN: 0-959455-1.

Glasson, Mikki and Glasson Ian " A Eucalypt Dyer's Handbook" in *Dyeing for Fibres and Fabrics,* Janet de Boer, ed. St. Lucia Queensland Australia: U. of Queensland ISBN: 0-959455-1.

Gohl, E.P.G. and Vilensky, L.D. 1980 2nd *edition Textile Science: An Explanation of Fibre Properties* Melbourne: Longman Cheshire ISBN:0-582-68595-8.

Goodwin, Jill 1982 *A Dyer's Manual* London: Pelham Books. ISBN: 0-7207-1327-7.

Goodwin, T.W. 1966 "The Carotenoids" in *Comparitive Phytochemistry* T. Swain, ed. London: Academic Press.

Goodwin, T.W. 1976 "Distribution of Carotenoids" in *Chemistry and Biochemistry of Plant Pigments vol. 1, 2nd ed.,* T.W. Goodwin, ed. London: Academic Press.

Goodwin, T.W. and Mercer, E.I. 1983 *Introduction to Plant Biochemistry, 2nd ed.* Oxford: Pergammon Press.

Griffiths, John 1976 *Colour and the Constitution of Organic Molecules* London: Academic Press. ISBN: 0-12-303550-3.

Hallett,Judith V. [1992] 1993 *Natural Plant Dyes,* Kenthurst NSW Australia: Kangaroo Press. ISBN: 0-86417-4381.

Harborne, J.B. 1966 "The Evolution of Flavonoid Pigments in Plants" in *Comparitive Phytochemistry* T. Swain, ed. London: Academic Press.

Harborne, J.B. 1967 *Comparitive Biochemistry of the Flavonoids* London: Academic Press.

Harborne, J.B. 1976 "Functions of Flavonoids in Plants" in *Chemistry and Biochemistry of Plant Pigments vol. 1, 2nd ed.,* T.W. Goodwin, ed. London: Academic Press.

Harborne, J.B. 1979 "Variation in Fuctional Significance of Phenolic Conjugation in Plants" in *Recent Advances in Phytochemistry vol. 12: Biochemistry of Plant Phenolics,* Tony Swain, Jeffrey B. Harborne, Chris F. van Sumere, eds. New York: Plenum ISBN: 0-306-40028-6.

Harborne, J.B. 1994 "Phenolics" in *Natural Products: The Chemistry and Biological Significance* J. Mann, R.S. Davidson, J.B. Hobbs, D.V. Banthorpe and J.B. Harbourne, eds. Essex England: Longman Scientific and Technical. ISBN: 0-582-06009-5.

Harborne, J.B and Williams, Christine A. 1982 "Flavone and Flavonol Glycosides" in *The Flavonoids: Advances in Research* J.B. Harborne and T.J. Mabry, eds. London: Chapman and Hall ISBN 0-412-22480-1.

Harris, Jennifer, ed. 1993 *Textiles 5000 Years: An International History and Illustrated Survey.* New York: H N Abrams.

Haslam, E. 1979 "Vegetable Tannins" in *Recent Advances in Phytochemistry vol. 12: Biochemistry of Plant Phenolics,* Tony Swain, Jeffrey B. Harborne, Chris F. van Sumere, eds. New York: Plenum ISBN: 0-306-40028-6.

Hendrickson, James B. 1965 *The Molecules of Nature: A Survey of the Biosynthesis and Chemistry of Natural Products* New York: WA Benjamin Inc.

Hess, Elizabeth *Natural Dyes* unpublished senior seminar paper, College of Mt. St. Joseph.

Heywood, H. 1966 "Phytochemistry and Taxonomy" in *Comparative Phytochemistry* T. Swain ed. London: Academic Press.

Hindmarsh, Lorna 1982 *A Notebook for Kenyan Dyers* Nairobi: National Museum of Kenya.

Hrazdina, Geza 1982 "Anthocyanins" in *The Flavonoids: Advances in Research* J.B. Harborne and T.J. Mabry, eds. London: Chapman and Hall ISBN 0-412-22480-1.

Hutchings, John B. 1998 "Color in Plants, Animals and Man" in *Color for Science, Art and Technology*, Kurt Nassau ed. Amsterdam: Elsevier ISBN:0-444-89846-8.

Isaacs, Jennifer 1984 "Indigo Textiles, Japan, Laos, Nigeria: Indigo in Nigeria" in *Dyeing for Fibre and Fabrics* Janet deBoer, ed. St. Lucia Queensland Australia: U. of Queensland ISBN: 0-959455-1.

King, R.A. 1999 "Role of Polyphenols in Human Health" in *Tannins in Livestock and Human Nutrition. ACIAR Proceedings No. 92* J.D. Brooker, ed. Canberra: Australian Centre for International Agricultural Research.

Kiracofe, Roderick, Johnson, Mary Elizabeth 1993. *The American Quilt: A History of Cloth and Comfort 1750-1950.* New York:Clarkson/Potters Pub. ISBN:0-517-57535-3.

Kraemer, Jack 1972 *Natural Dyes: Plants and Processes.* New York: Charles Scribner's Sons. *SBN (not ISBN)* 684-12828-4.

Krauskopf, John 1998 "Color Vision" in *Color for Science, Art and Technology*, Kurt Nassau ed. Amsterdam: Elsevier ISBN:0-444-89846-8.

Krohn, Val Friedling 1980 *Hawaii Dye Plants and Recipes* Honolulu: The U. of Hawaii Press.

Leistner, E. 1980 "Quinoid Pigments" in *Pigments in Plants, 2nd ed.* Czygan, F.-C. ed.Stuttgart: Gustav Fischer Pub.

Lesch, Alma 1970 *Vegetable Dyeing: 151 Color Recipes for Dyeing Yarns and Fabrics with Natural Materials.* New York: Watson-Guptill Pub. ISBN: 0-8230-5600-7.

Lewis, P.A. 1998 "Colorants: Organic and Inorganic Pigments" in *Color for Science, Art and Technology*, Kurt Nassau ed. Amsterdam: Elsevier ISBN:0-444-89846-8.

Liles, J. N. 1990 *The Art and Craft of Natural Dyeing: Traditional Recipes for Modern Use.* Knoxville TN USA: University of Tennessee Press. ISBN:0-87049-670-0.

Lloyd, Joyce [1971] 1978 *Dyes from Plants of Australia and New Zealand: A Practical Guide for Craftworkers.* Wellington New Zealand:AH & AW Reed. ISBN:0-589-01121-9. ISBN for the 1971 edition: 0-589-00658-4.

Maclaren, John A and Milligan, Brian (1981) *Wool Science: The Chemical Reactivity of the Wool Fibre* Marrickville NSW Australia: Science Press. ISBN 0-85583-091-3.

Martin, Vlada, Lamprell, Jeanne, Lucas, Shirley and Marks, Val eds. 1974 *Dyemaking with Australian Flora,* [1974] 1976 Sydney: Rigby Ltd. ISBN: 0-85179-664-8

Mayer, Fritz and Cook, A.H. 1943 *The Chemistry of Natural Coloring Matters,* ACS Monograph Series, New York: Reinhold Pub. Co.

McGregor, R. *Textile Research Journal* 42 (1972) 536-552.

Mead, Hirini Moko 1999 *Te Whatu T niko: T niko Weaving Technique and Tradition.* Aukland: Reed Books. ISBN:0-7900-0679-0.

Milner, Ann [1971] 1980 *Natural Wool Dyes and Recipes* Dunedin New Zealand: John McIndoe.

Motomura Hiromi, Bae,Sook-Hee and Morita Zenzo *Dyes and Pigments* 39 (1998) 243-258.

Myers *Navajo Dye Chart*

Nassau, Kurt 1998a "Fundamentals of Color Science" in *Color for Science, Art and Technology,* Kurt Nassau ed. Amsterdam: Elsevier ISBN:0-444-89846-8.

Nassau, Kurt 1998b "The Fifteen Causes of Color" in *Color for Science, Art and Technology,* Kurt Nassau ed. Amsterdam: Elsevier ISBN:0-444-89846-8.

Nassau, Kurt 1998c "Color Preservation" in *Color for Science, Art and Technology,* Kurt Nassau ed. Amsterdam: Elsevier ISBN:0-444-89846-8.

Navajo School of Indian Basketry [1903] 1971 *Indian Basket Weaving* New York: Dover reprint. ISBN:0-486-22616-6.

Neilsen, Edith 1984 "Drying for Dyeing" in *Dyeing for Fibre and Fabrics* Janet deBoer, ed. St. Lucia Queensland Australia: U. of Queensland ISBN: 0-959455-1.

Padgett, Rosa 1964 *Textile Chemistry and Testing in the Laboratory* Minneapolis MN USA: Burgess Pub. Co.

Ponting, K.G. 1980 *A Dictionary of Dyes and Dyeing* London: Mills and Boon Ltd. ISBN: 0-263-06398-4.

Pyott, Steven 1985 *Textile Care and the Consumer* Melbourne:Longman Cheshire Pty. Ltd. ISBN: 0-582-87233-2.

Ramsey, Bets and Waldvogel, Merikay 1998 *Southern Quilts: Surviving Relics of the Civil War.* Nashville TN: Rutledge Hill Press.

Robinson, Stuart 1969 *A History of Dyed Textiles* Cambridge MA USA: MIT Press.

Robinson, Trevor 1967 *The Organic Constituents of Higher Plants, 2nd ed.* Minneapolis MN USA: Burgess Pub. Co.

Ronsheim, Ellice 1991 "From Bolt to Bed: Quilts in Context" in *Quilts in Community: Ohio's Tradition,* Ricky Clark, George W. Knepper and Ellice Ronsheim eds. Nashville TN: Rutledge Hill Press.

Sayer, Chlöe 2002 *Textiles from Mexico* London: The British Museum Press ISBN: 0-7141-2562-8.

Schetsky, Ethel-Jane McD., Woodward, Carol H., Scholtz, Elizabeth eds. 1964. *Dye Plants and Dyeing: A Handbook.* New York: Brooklyn Botanical Gardens

Sequin-Frey, M. *J. Chem. Educ.* 58 (1981) 301-05.

Shoemaker, Garland, Nibler 1989 *Experiments in Physical Chemistry, 5th ed.* New York: McGraw Hill

Sime, Rodney J. 1990 *Physical Chemistry: Methods, Techniques and Experiments* Philadelphia: Saunders College Publishing

Simon, Frederick T. 1980 "Color Order" in *Optical Radiation Measurements vol. 2: Color Measurement* Franc Grum, and C. James Bartleson, eds. New York: Academic Press ISBN: 0-12-304902-4.

Smith, Alick 1984 "Vegetable Dyeing" in *Dyeing for Fibre and Fabrics* Janet deBoer, ed. St. Lucia Queensland Australia: U. of Queensland ISBN: 0-959455-1.

Storey, Joyce 1978 *The Thames and Hudson Manual of Dyes and Fabrics* London: Thames and Hudson.

Swain, T. 1976 "Nature and Properties of Flavonoids" in *Chemistry and Biochemistry of Plant Pigments vol. 1, 2nd ed.,* T.W. Goodwin, ed. London: Academic Press.

Swain, Tony 1979 "Phenolics in the Environment" in *Recent Advances in Phytochemistry vol. 12: Biochemistry of Plant Phenolics,* Tony Swain, Jeffrey B. Harborne, Chris F. van Sumere, eds. New York: Plenum ISBN: 0-306-40028-6.

Swain, Tony 1980 "Flavonoids as Chemotaxonomic Markers in Plants" in *Pigments in Plants, 2nd ed.* F.-C. Czygan, ed.Stuttgart: Gustav Fischer Pub.

Thomson, R.H. 1976 "Quinoids: Nature Distribution and Biosynthesis" in *Chemistry and Biochemistry of Plant Pigments vol. 1, 2nd ed.,* T.W. Goodwin, ed. London: Academic Press.

Thomson, Ronald H. 1978 "Quinoid Compounds" in *Natural Compounds Part 3 Steroids, Terpenes and Alkaloids,* F. Korte and M. Goto eds. New York: Academic Press ISBN: 0-12-460747-0.

Thompson, Frances and Thompson, Tony 1987 *Synthetic Dyeing for Spinners, Weavers Knitters and Embroiderers.* Melbourne: Nelson Pub. ISBN:0-17-007192-8.

Toguchi, Mary 1984 "Indigo Textiles, Japan, Laos, Nigeria: Indigo in Japan" in *Dyeing for Fibre and Fabrics* Janet deBoer, ed. St. Lucia Queensland Australia: U. of Queensland ISBN: 0-959455-1.

Trotman, E.R. 1970 *Dyeing and Chemical Technology of Textile Fibers, 6th ed.* High Wycombe, England: Charles Griffen & Co. Ltd.

van Stralen, Trudy 1993 *Indigo, Madder and Marigold: A Portfolio of Colors from Natural Dyes.* Loveland CO: Interweave Press. ISBN: 0-934026-86-6.

Waldvogel, Merikay 1990 *Soft Covers for Hard Times: Quiltmaking and the Great Depression* Nashville TN: Rutledge Hill Press.

Waldvogel, Merikay 1994 "Southern Linsey Quilts" in *Quiltmaking in America: Beyond the Myths,* Laurel Horton, ed. Nashville TN: Rutledge Hill Press.

Waterman, P.G. 1999 "The Tannins – An Overview" in *Tannins in Livestock and Human Nutrition. ACIAR Proceedings No. 92* J.D. Brooker, ed. Canberra: Australian Centre for International Agricultural Research.

Welch, M.B. and Coombs, F.A. 1926 *The Principal Tanning Materials of Australia and their Leather Forming Properties. Bulletin No. 10* Sydney: Technology Museum.

Were, Ian ed. 2003 *Story Place: Indigenous Art of Cape York and the Rainforest* Brisbane: Queensland Art Gallery. ISBN:

Wickens, Hetty 1983 *Natural Dyes for Spinners and Weavers* London: BT Batsford Ltd. ISBN: 0-7134-4228-X.

Wiegle, Palmy 1974 *Ancient Dyes for Modern Weavers* New York: Watson-Guptill ISBN:0-8230-0223-3.

Wingate, Isabel B. 1976 *Textile Fabrics and Their Selection, 7th ed.* Englewood Cliffs NJ USA: Prentice Hall. ISBN:0-13-912840-9.

Wright, Dorothy 1978 *The Complete Book of Baskets and Basketry* Sydney: AH & AW Reed ISBN:0-589-01115-4.